ADAM

DƎCŌDƎD

A BRIEF HISTORY OF MAN'S

TRUE ORIGINS

VOL. 2

⭐ LEON BIBI ⭐

Pegasus Publishing Company

Printed in the United States of America
First Printing 2018
First Edition 2018

10 9 8 7 6 5 4 3 2 1

ISBN 13 978-1720247777

Library of Congress
Control Number: 2018956611

"The Lord God said, Behold the man
is become

as *one of us*"

(Old Testament 3:22)

Warning:

This book contains a secret, the topics of which have been forbidden from human history for 4,000 years.

𒀭𒈠𒅔𒍑𒐅𒍑𒐕𒌋

DEDICATION

To My Father: Morris Bibi who died at age 94 having fulfilled a long and fruitful life.

and

To Alan Alford: Who died unexpectedly in Nepal. Alan's book - "Gods of the New Millennium" was a monumental breakthrough in the ancient alien theory.

TABLE OF CONTENTS

ADAM DƎCŌDƎD

TIMELINE

DATE	EVENT
443,000 BC -	Arrival of the Anunnaki
442,000 BC -	360,000 BC - Anu comes to Earth
415,000 BC -	Ninhursag established her Medical center in Shurupak
335,000 BC -	Ice Age
226,983 BC -	Enki moves to Africa to supervise mining
220,000 BC -	Ice Age
183,783 BC -	Rebellion of the Anunnaki mine workers
180,000 BC -	Creation of Homo Sapiens
176,583 -	Garden of "E.DIN" - Adam and Eve procreate
127,000 BC -	Ice Age
20,983 BC -	Noah is born with a "genetically engineered skin color"
20,880 BC -	Noah has 3 children- Shem, Ham and Japheth
12,364 BC -	Nibiru enters Earth's orbit and forces Earth to tilt on axis
11,600 BC -	Ice Age
10,983 BC -	The Flood - Antarctic Ice Sheet slips into Indian Ocean. Massive tsunami overwhelms the Arabian Peninsula and floods the Persian Gulf
10,450 BC -	The Pyramids of Giza are built
8,764 BC -	Nibiru returns to Earth's orbit
8,700 BC -	Jerusalem built as a space facility. The Sphinx is carved
8,670 BC -	Second Pyramid War ends
5,164 BC -	Nibiru returns to Earth's orbit
4000 BC -	Uruk becomes the largest city in Mesopotamia. Anu visits Earth again.
3800 BC -	Sumerians write down first recorded history on clay tablets, in Uruk
3760 BC -	The Jewish calendar begins
3450 BC -	Nimrod builds the Tower of Babel for Marduk, and it is destroyed by Enlil

TIMELINE

DATE	EVENT
3100 BC -	Pharaonic dynasties begin
3000 BC -	Stonehenge is built by Thoth for Marduk as a star-clock
2700 BC -	Enlil resides in Nippur
2500 BC -	Uruk has a population of 40,000 people
2123 BC -	Birth of Abraham
2024 BC -	Anunnaki set off a nuclear weapon at Sodom and Gomorrah The Dead Sea bears its name. Many Anunnaki leave the earth
2024 BC -	Famine and hardship from the blast ends Sumerian civilization
2001 BC -	Jacob is born
1992 BC -	Abraham dies
1513 BC -	Moses discovered in the river
1450 BC -	Mohenjo-Daro and Harappa in current day Afghanistan destroyed by a nuclear bomb
1393 BC -	Yahweh created by Israelites as the one true God is documented in the Old Testament
1391 BC-	Teotihuacan pyramids are built
1308 BC -	Israelites exodus out of Egypt
968 BC -	King Solomon born
946 BC -	King Solomon builds Temple of Jerusalem for Yahweh
925 BC -	Temple of Jerusalem is destroyed by Ramses the Great
610 BC -	The balance of the Anunnaki left Earth
586 BC -	Nebuchadnezzar burns King Solomon Temple to Enlil
500 BC -	Old Testament was written
200 BC -	Anunnaki leave Earth for good

PREFACE

"We live in a world of paradox, of great duality. We have taller buildings but shorter tempers. Wider freeways but never viewpoints. We retain more information but have less wisdom. We have more options for leisure and less fun. More conveniences, yet less time. We have more medicine and medical options, yet less wellness. More kinds of food available, but less nutrition. We have more media outlets but less communication. More acquaintances, but fewer friends. We have increased our positions but reduced our values. We have conquered outer space but neglected inner space. We have smashed the atom, but not our prejudices." - Author Brad Olsen

"Search for yourself, by yourself. Do not allow others to make your path for you. It is your road and yours alone. Others may walk it with you, but no one can walk it for you." - Native American Code of Ethics

"Are some of us guilty of bias when we treat 5,000-year-old clay tablets as myth, but the 2,500-year-old Genesis text as fact?" - Alan Alford

THIS BOOK, MY SECOND ON THE SUBJECT of human origins, is an attempt to convince you, my dear reader, that we are in fact in the midst of a truth paradigm shift. More and more evidence is coming to the surface to prove that we must reconsider and rethink the origins of man and the powers that be that protect a lie. The "missing link" will be eternally missing. There is no truth to perfect Darwinism, and the colossal evidence that exists in the form of monumental structures, cuneiform tablets and written texts tell the true story of human evolution. We are not, and have never been the only sentient, living, breathing humanoid being in our galaxy, or the universe. We are infants living in a world of mature adults that are only now disciplining us on how to behave. The trouble is, we don't want to behave.

When Einstein got frustrated with the behavior of light in the framework of the known Newtonian physics, he created instead a *new concept* of physical reality - a paradigm shift - which he coined the Theory of Relativity. In order to solve the quandary of human evolution, we cannot solve the problem with the same kind of thinking that gave rise to the problem. We must think outside the box and delve into the uncomfortable realms of our deepest fears. Our deepest, darkest truths that we know, deep down inside, are real. The evidence of UFO's and extraterrestrial existence is overwhelming. It exists right under our noses, but our nascent, childish fears kick in to subdue the evidence and live comfortably numb.

The sciences of quantum physics, cosmology and evolutionary biology are at the very core of this shift. They are all interrelated and unified in nature. The universe is 13.7 billion years old. Earth is only 4.6 billion years old - leaving a whopping 9.1 billion years for other intelligent species to evolve. Now there are two camps that theorize how the universe was created - the proponents of intelligent design, and the proponents of straight evolution. Maybe, in the words of physicist Ervin Laszlo, it is *"design for evolution"* (Laszlo - Science and the Akashic Field). Maybe this is true. Everything was designed on purpose to evolve. If Harvard astronomer Harlow Shapley is right, then there are at least 100,000,000 planets capable of supporting life in the cosmos. Based on the Drake equation alone, there are 10,000 advanced technological civilizations that are likely to exist in our Milky Way galaxy alone!

EVIDENCE - our evidence of mankind's true origins is revealed by the mysterious fusion of our chromosomes and extreme increase in our brain capacity in short time.

MOTIVE - the motive of our makers was to create a slave race to carry out physical labor in mines in Africa and South America to excavate gold, silver and other minerals necessary to save the Anunnaki atmosphere from incinerating.

BARRIERS - the barriers that exist to prevent the "truth" from being told to our children are religion and government.

- religion is preventing digging underneath the Sphinx
- religion is preventing digging in Jerusalem
- religion is to blame for the destruction of millions of pages of history from the library in Alexandria, to the burning of "works of the devil" in Europe and the Americas by missionaries.
- the powerful religious families worldwide *know* the truth, but do not want to leak it, because it will cause religion to crumble upon its foundations

CHAPTER 1

MYTH or HISTORY?

⟨⟨ ⍦ ⋉⋀ ⊐⍦ ⋸⋪ ⊐⋀ ⋁⊀ ?

"There no laws of physics, astronomy or biology that can rule out extraterrestrial involvement in the evolution of life on Earth... Positing that an advanced race came from a distant solar system, seeded life on Earth, then intervened in human affairs at infrequent intervals does not conflict with the Old Testament at all, it agrees with it. How will they react to the theory that, while evolution exists, it did not start on planet Earth, and the reason there are no missing links is (that) none ever existed" - Will Hart

ISN'T IT INTERESTING THAT IN school anything we learn as "History" is only that which is supported by the scriptures - the Old Testament and the New Testament. History is considered nonfiction and supported on the basis of actual fact. Anything in school that we learn as "Myth" is only that which is not supported by the scriptures. Myth is considered fiction and not supported on the basis of actual fact. What if I told you that the exact opposite was true? What if I told you that myth does have basis in actual, proven fact, and much of history is an absolute farce - a cover-up, that has no proven basis in fact. Why should the scriptures be the basis of that which is fact?

Scriptures should only be the basis of faith, not fact. Only tested, then re-tested science can be considered proven fact. Much of the scriptures cannot be proven at all, and a large part of the scriptures can be considered lunacy.

Consider the following biblical stories -

- ✓ Noah transporting every living species of animal upon his ark to escape the flood
- ✓ Christ walking on water and Reviving the Dead
- ✓ Moses parting the Red Sea and then releasing it to kill thousands of Egyptians
- ✓ The Jews surviving for 40 years in the desert on Mana alone

"Why is it, then, that so many of those same scholars uphold the Church's veneration of Genesis as an absolute truth, whereas they decry the original records as legend and mythology? It is because, in the final analysis, despite failing congregations, Church opinion always wins at an official level since it is inherently tied to the governments which control the academic establishments." (Gardner - pg. 83)

Now, since classroom textbooks are based upon the chronology of the scriptures, it is my opinion that their chronology is incorrect. It is not true that the Civilized World began in 4004 BC as was told by Archbishop Ussher. We have unearthed civilizations in Turkey such as

Gobekli Tepe which has been carbon dated to before 10,000 BC. My previous book **"Adam = Alien"**, also States that the Pyramids of Giza and the Sphinx maybe dated to as far back as 10,000 BC as well, based upon proof of water erosion at their base. If this is true then the current chronology of the Egyptian pharaohs must then be incorrect. As Gardner states in *Genesis of the Grail Kings* -

"Therefore, when certain Pharaohs were identified (correctly or incorrectly) as being the unnamed or loosely named pharaohs of the Bible text, their dates were plotted in accordance with the standard Old Testament reckoning. Then by counting the regnal years backwards and forwards from the strategic points the Egyptian chronology that we now have in our authorized textbooks was constructed. This pharaonic chronology is entirely dependent on the presumption that the standard biblical chronology is correct - but the Bible chronology of Archbishop Ussher and the Christian church is far from correct...Where Sumerian history is concerned, we are looking at texts with much older Roots than the earliest Egyptian records so far discovered...We are told that our children are being shielded from the romance of mythology, but they are actually being prevented from learning the truth of history. This is a purposeful, strategic manipulation by an establishment which knows only too well that **learned people are the greatest of all threats to governmental thralldom.** *"* (Gardner - pg. 91)

Based upon the famous Sumerian tablet - *The Kings List* - it can be deduced that the eight pre-flood Kings reigned for a total of 24,120 years. Gardner is of the belief that the flood occurred in 4000 BC, he then believes that civilized man originated in 30,000 BC. It is my belief however, that the flood occurred in 12,000 BC, pushing back man's earliest civilization to 36,000 BC.

SUMERIAN TABLET

THE GREAT FLOOD

The Great Flood as we know it was, before the Bible, originally described in the *Epic of Gilgamesh* - one of the Sumerian tablets found and transcribed by Leonard W. King in his book - *"The Seven Tablets of Creation"* in 1902. In it he described a variety of gods, not just one as in the Bible. It must be stated again that the cuneiform tablets pre-date the Bible by thousands of years. Even earlier, a young banknote engraver and amateur Assyriologist named George Smith who worked for the British Museum in London had been assembling the tablets and noted that their translation and stories were very similar to our Bible's description of the Great Flood. His book *"The Chaldean Account of Genesis"* was written in 1876 and was the first to compare the ancient texts discovered in Mesopotamia with the Creation tales and Flood tale in the Bible. In it, the Flood had been described **firsthand** by its author - Atra-Hasis - an actual man. Was this Noah speaking directly to us through the tablets about the Flood? According to Sitchin's research of the list of pre-Flood rulers, it was **indeed**, and it was none other than the 10th direct decedent of Adam!

Here is an exact transcription of the Flood tablet in which Enki reveals to Noah the coming of the Flood:

Man of Shuruppak, son of Ubar-Tutu:

Tear down the house, build a ship!

Give up possessions, seek thou life!

Forswear belongings, keep soul alive!

Aboard ship take thou <u>seed of all living things</u>.

That ship thou shalt build-

Its dimensions shall be to measure.

Genesis 8:4 states that Noah's Ark landed in the mountain range of Ararat which was just to the north of Mesopotamia, specifically the mountain of Lubar. My previous book asserts that the great flood was caused by Enlil who was furious with humans and the noise that they caused. After hearing in the great assembly of the Anunnaki that Enlil was going to cause the great flood, Enki warned Noah about this. Noah then built an enclosed submersible ship, not an open great ark, which would convey "*the seed of all living creatures*". It is my belief that this Arc contained the seeds of human and animal life in the form of DNA, not the actual animals themselves. It is more logical to assume that based upon the level of scientific advancement that the Anunnaki possessed, Enki taught Noah how to use DNA in the form of test tubes in order to transport the essence of each animal species. Noah was 600 years old when the flood swept over the Earth.

It is interesting to note that while the flood was essentially a colossal tidal wave caused by the collapse of the Ice Sheet over Antarctica, it caused devastation of the gold-mining facilities in South Africa. However, in South America, or the opposite side of the globe, the avalanche of water exposed gold in the Andes mountains in Peru and made it easy for the Anunnaki to obtain it without the need for mining.

Zecharia Sitchin (July 11, 1920 – October 9, 2010) is an author and historian renowned for his work in the Ancient Astronaut space, Sumerian mythology and analysis, and Egyptology. He was an innovative and detail-oriented investigator of pre-history human culture and origins. His books (example - "The 12th Planet") sold millions of copies and has been translated in 25 languages. Sitchin's work was tireless and unselfish, and he was not driven by greed or ego, rather by discovering the truth. His writing has been deeply inspiring to me.

SUMERIAN TABLET DESCRIBING THE FLOOD

THE OLD TESTAMENT

Prior to the Old Testament which identified God in the singular, monotheistic version, the plural concept of the Elohim gods was for the most part ignored. This is evidence to support the fact that the Old Testament was largely founded upon Sumerian lore. Gardner posits:

*"...it must be remembered that the 19 inclusive generations from Adam to Abraham were natives of Mesopotamia: Therefore, when Abraham migrated to Canaan in about 1900 BC, he arrived neither as a Jew, nor as a Canaanite, **but as a Sumerian.** He was, nonetheless, the first of the succession to be formally classified as Hebrew and he's regarded as the ultimate patriarch of the Jewish race. This stems from his Covenant with Jehovah - or more correctly with El Elyon. Henceforth, Abraham became the designated father of his people and male circumcision was adopted by his descendants."* (pg.36)

He continues-

"The name Hebrew derives from the patriarch Eber (Heber / Abhar,) six Generations before Abraham. The term 'Israelite' comes from the renaming of Abraham's grandson Jacob, who became known as Israel (Genesis 35:10-12). By way of translation, Is-ra-el means 'soldier of El'...As for the word 'Jew,' this comes from the style Judean which are the Hebrew Israelites of Judea in Canaan" (pg. 36)

Just as the word 'Bible' comes from the Greek plural noun *biblia* meaning 'a collection of books', the Old Testament can also be seen as a collection of several documents. In fact, even at the time of Jesus, there was no one singular composite Old Testament available to Jews at all. There were only various books and chronologies. This is proven by the 38 Scrolls of the 19 Old Testament books found at Qumran, Israel between 1947 and 1951.

THE PLAGUES

A possibility held by Dr. John Marr and Curtis Malloy regarding the scientific explanation of some of the ten plagues is that the changing of the color of the river Nile from blue to red had been due to-

1. An algae called Pfiesteria, causing the red discoloration and killing the fish (Blood)
2. Frog population increased because fish stopped eating eggs laid by the frogs (Toads)
3. Frogs left the river and died on land
4. Insects swarmed and fed off the dead frogs (Locusts)

RED ALGAE PFIESTERIA IN NILE RIVER

SODOM AND GOMORRAH

I always wondered what exactly happened to Lot's wife when she turned around during the destruction of Sodom and Gomorrah and was turned into a *"pillar of salt"*. Why a *"pillar of salt"*? Why not just killed? Genesis states that "sulphurous fire" came down from heaven and that the Lord "cast a thunderbolt upon the city and set it on fire with its inhabitants". According to R.A. Boulay in "Serpents and Dragons" -

"In the Haggadah, this thunderbolt comes from the *Shekinah,* which had *descended* to work the destruction of the cities." and "Lot and his family had been warned not to look behind them lest they be blinded by the flash of the explosion, **probably nuclear in nature**".

Now, why say that Lot's wife ought not to look behind her? Why not just say that Sodom and Gomorrah were just pulverized by some gigantic blast administered by God? Two key word exist here that should be discussed - *"blinded"* and *"pillar of salt"*. Blinded because the blast was so intense that it damaged the retina, and *"pillar of salt"* is actually a Hebrew word meaning *"pillar of gas"* or vaporized. She was vaporized by a nuclear blast. Doesn't it make sense? The blast was so intense that it actually created the Dead Sea. The Dead Sea is "*dead*" because the nuclear blast destroyed everything living in it.

SODOM AND GOMORRAH

LEVITATION

"If you want to find the secrets of the universe, think in terms of energy, frequency and vibration." - Nikola Tesla

In "Gods of Eden", Andrew Collins discusses an interesting case of a Swedish doctor named only 'Jarl' who in the 1920's or 1930's (date not specified) visited a monastery in Lhasa, Tibet. Here he was witness to an incredible event. This event may hold the key to how the pyramids and other monoliths built with heavy stones were actually constructed with such ease and sophistication.

The objective of the monks was to lift stone blocks up 250 meters above the side of a mountain wall to insert them by a cave up the cliffside. The setup was as follows:

1. 200 monks
2. 13 drums
3. 6 trumpets

Each instrument was placed 5 degrees apart to form the angle of an arc of 90 degrees (similar to a pie shape). 10 monks stood behind each instrument. One monk at the center of the "pie" chants in a low monotone voice and hits the drum with one hand. Then the trumpets and other drums followed suit. No one spoke other than the monk in the middle. After four minutes, the stone block began to wobble! Then it lifted into the air and rocked side to side. As the trumpets and drums tilted, the rock rose upwards. It rose higher and higher until it reached the top of the cliff. Absolutely incredible! Jarl was stunned. An analogy for this process would be the opera singer that can shatter glass with her voice. Her vocal chords produce a sound that "exaggerate the inherent vibrational frequency of the glass. This has the effect of causing it to oscillate or shake so violently that eventually it disintegrates".- Gods of Eden – Collins

John Ernst Worrell Keely (1827-1898), an American from Philadelphia may be the master of unlocking the mechanics of sound levitation. Between 1880 and 1892, Keely devised a system using 4 key components:

1. Liberator – the 'master' ingredient which operated the other 3 components. Keely would strike this first, then tune it
2. Resonator- had a compass that spun
3. Musical Note- a harmonicon trumpet that Keely played for 2 minutes
4. Connecting Chord- connected to the "globe" that would spin

Keely then suspended 2-lb spheres on top of water. (Author's note – was this how Jesus supposedly walked on water?). The concept was that "sympathetic vibration" could disintegrate quartz and other hard rock. Keely also used gongs, horns, harmonicas and fiddles.

Was the infamous "ram's horn" or "shofar" of the Israelites during their movements throughout the Middle East used for the same purpose? As a weapon against their enemies?

During the infamous Anunnaki mining in South Africa, Enki combined the following 8 precious metals - ruthenium, rhodium, palladium, silver, osmium, iridium, platinum and gold to create a *transition group* of elements. He knew that by combining these elements they would, in a monoatomic low energy state, become superconductive, and create levitation.

LEVITATING MONK

CHAPTER 2

ANCIENT MACHINERY

THE PYRAMIDS

BELIEVE IT OR NOT, THERE ARE 600 pyramids all shaped in the proportion of *pi,* on Earth. There are even 2 underwater near Japan and Cuba built 12,000 years ago!

Every Egyptologist you will read about or see on television will tell you that the pyramids were built as tombs for pharaohs. They will cite all the pharaohs that have ruled Egypt for millennia, and all the reasons why each pharaoh was buried in a particular pyramid in a particular city. But here's the catch - they can't prove it!

Where are the mummies that were supposedly buried in these pyramids? Yes, they have uncovered mummies - Tutankhamen, etc. But they weren't buried in pyramids. They were buried in tombs deep underground. So why were the pyramids built, then? Why spend so much time and (massive) effort to construct these colossal monoliths? There has to be a better answer.

Why were the ancient Egyptians obsessed with the precession of the equinoxes? Was this only so that they could predict the weather patterns in order to cultivate their crops at the right times? Why make shafts that run from the King's and Queen's chambers to the outside? Why was Rudolf Gantenbrink, who developed a robot that ascended the southern shaft of the Queen's chamber to find a door equipped with two copper fittings, fired the week after his discovery? One would think that such a discovery would be cause for celebration and further exploration of this door, and especially the purpose of the copper fittings. Why copper? Where is the capstone (top triangular part) today? Was it originally a fully functional giant crystal of the Giza Pyramid which could distribute energy? Like a prism?

THE GREAT PYRAMID OF GIZA

Five reasons why the pyramids could not have been built 5,000 years ago by humans:

1. The scale of construction was colossal

2. The building materials (limestone and granite) are the heaviest and most difficult materials to manipulate and shape, and no cutting tools for granite had existed in 2000 BC. Copper tools that were found near the Great Pyramid site were, and are, incapable of cutting granite

3. The precision of the pyramids with exact mathematical equations hadn't existed 5,000 years ago

4. There are machine tool marks that hadn't existed 5,000 years ago

5. The base of the Great pyramid, after having been surveyed using modern instruments, was found to be exactly level to within ⅞ inch

This fourth point is the death blow to Egyptologists that continue to rehash old, tired and nonsensical rhetoric detailing Jewish slaves laboring to build the pyramids.

Taking it a bit further, author Brad Olsen cites the following:

"The Great Pyramid is the most fascinating of all pyramids worldwide, located exactly at the center of all land masses on Earth. Such precise global configurations could only be observed from an aerial perspective above Earth, or from outer space. The ratio between the height of the Great Pyramid and its perimeter is the same as the

ratio between the earth and its circumference. The exact measurements and the fact that there are doors within the air shafts suggest a machine-like function. Charting the mathematical calculations of the geodetic center of the earth continents could not be made in any other way except with utilizing advanced airplanes or spacecrafts. " - Olsen - Modern Esoteric - page 170

Another point that needs to be addressed is that *newer* pyramids built during the Fourth Dynasty – approximately 100 of these – are now in a state of deterioration. You must ask yourself the following – if in fact the older pyramids (Giza) were built 5,000 years and before the Fourth Dynasty pyramids, why would they be built with sub-standard construction? Why would the Egyptians not use their supposedly miraculous techniques again? It makes no sense. I believe that the Giza pyramids were constructed with the direct help of the Anunnaki using power saws and drills, and that the later pyramids of the Fourth Dynasty were attempts by humans to copy the pyramids *without* help.

Two million cubic meters (seventy million cubic feet) of limestone was cut to build the Great Pyramid. More stone was cut building the pyramids of Giza than was used in all building for 1500 years from 1550 BCE to 30 BCE!

Examples of pyramid construction positioning, *in all of the earth,* infers that the builders *had to* build them in certain locations can be seen in the Second (Khafra's) Pyramid. The plateau had been sloping and needed to be leveled. The builders had to create *steps* in the rocks and *rises* to create the lower part of the slope. The builders could have built the Second Pyramid just a few hundred meters away from this position which did, in fact, have level ground....but they did not choose that spot. If the builder wanted ease of production, they wouldn't have built the Second Pyramid in the spot where it rests today.

Chris Dunn (discussed later) suggests that the builders used high-powered drills which spun 500 times faster than modern boring drills with diamond tips! Dunn proposes that they used "ultrasonic drills' which use sound to make the bit vibrate at an enormously high rate"- "The Stargate Conspiracy" – Picknett and Prince

The authors Picknett and Prince also point out that the CIA got heavily involved with author Robert Temple after his book *"The Sirius Mystery"* was published in 1976. They harassed him constantly, probing him about how he received the information that he published. Why? Temple claims that he was harassed for 15 years! He found out that MI5 carried out security checks on him and that the British Security Services commissioned a report on the book. He was also approached by the American Freemason Charles E. Webber, who asked him to join the Freemasons to become a "33rd Degree Mason" – the highest rank in Freemasonry. Webber told him-

"We are very interested in your book The Sirius Mystery. We realize you have written this without any knowledge of the traditions of Masonry, and you may not be aware of this, but you have made some discoveries which relate to the most central traditions at a high

level, including some things that none of us ever knew" – "The Stargate Conspiracy" – Picknett and Prince

Under the Sphinx, radar has detected 9 concealed chambers of manmade origin. It has yet been touched. Could this be the "Hall of Records" that discloses our true identity?

PAWS OF THE SPHINX

DOES THIS STONE REPRESENT THE HIDDEN TUNNEL LEADING TO THE "BOOK OF KNOWLEDGE"?

Evidence of machining – in the King's Chamber – on the red granite sarcophagus, there are horizontal and vertical saw marks. Evidence exists that the saw marks were started incorrectly, then re-started to correct the cuts. It had been backed out, repositioned, then restarted again. Additionally, the surface was polished to hide the mistake! THIS IS INCREDIBLE EVIDENCE that the Anunnaki had used saws approximating 9 FEET LONG to cut the hard granite to shape the sarcophagus! They made two mistakes and started over. The bronze knives that were found 2500 BC could NEVER have accomplished what these saws accomplished.

Now, humans could have used sand-based abrasives to cut through alabaster and limestone, but not hard igneous rocks such as dark basalt, red granite and black diorite. Impossible!

The saws used were probably made of bronze with diamond-tipped teeth. This is the only substance that could have cut cleanly through this igneous rock.

Drills and lathes were also used, and there is ample evidence of their use.

These tools could not have been used by humans at that time.

Sir William Matthew Flinders Petrie (1853-1942), an English Egyptologist, summed up the drilling process as follows – (taken from "Gods of Eden" – Andrew Collins):

1. The granite cores produced by tubular drilling seemed to taper towards the top, i.e. at the point where the drill entered the stone, while the circular wall of the borehole always appeared to be wider at the top
2. The jeweled drill-piece left behind perfect grooves that swirled around the circumference to form a regular, symmetrical spiral without waviness or interruption; in one case 'a groove can be traced, with scarcely an interruption for a length of four turns'.
3. The (spiraling) grooves are as deep in the quartz as in the adjacent feldspar, and even rather deeper

Christopher Dunn, an American tool engineer par excellence, who has written extensively about tool machining in Egypt expressed that the machine workers were "able to cut their granite with a feed rate that was *"500 times greater"* than what we can produce today! Can this be believed? That the pyramid builders of 4500 years ago could cut granite at a speed 500 times greater than our best today? What more evidence does one need to prove that advanced Anunnaki builders with huge saws cut and shaped the pyramids. They used a process known today as "ultrasonic drilling" which means they used a high-pitched inaudible sound to cause the drill-bit to vibrate at incredibly fast speeds.

John Anthony-West discussed the extreme civilization boost in Egypt with the following-

"How does a complex civilization spring full-blown into being? Look at a 1905 automobile and compare it to a modern one. There

is no mistaking the process of 'development'. But in Egypt there are no parallels. Everything is there right at the start'. It is rather as if the first motor car was a modern Rolls-Royce." – Colin Wilson – "From Atlantis to the Sphinx" (pg. 17)

Why was the star Sirius so important during the building of the Pyramids? It rose at dawn at the beginning of the Egyptian New Year when the Nile began to rise.

Robert Bauval, a best-selling author on Egyptology, has an interesting concept based upon his date of 10,450 BC being the exact alignment of the 3 pyramids with the Belt of Orion that this alignment may happen to coincide with the fall of Atlantis. It may have been a "rebirth" after the destruction of Atlantis (which I believe took place in Antarctica).

Another connection between Egypt and Atlantis lies with the large ships found buried in Egypt. It is a mystery as to why the Egyptians would build a 143-foot ship to sail on the Nile? Seems way too big. John Anthony-West and Schwaller de Lubicz (another original and creative Egyptologist) believe that the ships were solely ritual objects intended to glorify the Atlanteans who came to Egypt after the destruction of Atlantis *in ships*!

Madame Blavatsky, an eminent psychic and author of the late 19th century, wrote that current humans live in what she calls the "Fifth Root Race". She claims that the Fourth Root Race was the Atlanteans in Antarctica, the Third was the Lemurians in Australia, the Second on the continent of Mu, but she fails to mention who the First Root Race was. I believe it was the original Anunnaki offspring in Africa – Adamu. So, we would effectively be the Fifth Generation of homo sapiens offspring of the original hybrids.

Another interesting point regarding pyramid engineering is that the inside of the Sun Pyramid contains a layer of mica. Why mica? Not only does it contain this layer of mica, the mica is from far away. We discussed the fact that many building materials for the 3 pyramids came from hundreds of miles down the Nile River, however, this mica came from *thousands of miles* away this time – *from Brazil!* Why would they do this? --According to Colin Wilson in "From Atlantis to the Sphinx."

"(and) how were 90-foot sheets of mica transported? Moreover, why was it then placed under the floor? What purpose did it serve there? Graham Hancock points out that the mica is used as an insulator in condensers and that it can be used to slow down nuclear reactions…" (pg. 129)

How interesting! This all loops back to my original belief that the Pyramids were in fact used as an energy tower. It makes sense. If the pyramids were used as tombs for dead Kings, why use 90-foot sheets of mica from two thousand miles away in Brazil? For decoration? Makes no sense.

Could it be possible that the Sphinx was built first in 10,500 BC – the year that Orion would reach its lowest point in the sky – and then the pyramids were built 8,000 years later, in 2,500 BC when a "new age" began?

Could it be possible that there lies a secret tunnel, 700 meters long – under the Sphinx, leading to the Great Pyramid? Apparently, only the Egyptian Antiquities Authority is permitted to dig beneath the Sphinx and may already have done so. We have no information about what they are doing or what they have found so far. The famous psychic Edgar Cayce said that there was a secret room beneath the rear paws of the Sphinx.

There is proof that the Pyramids were at one time, within close proximity to a body of water. There is a full fossilized whale skeleton from the Eocene epoch found near Fayum. This 50-foot specimen dubbed "Basilosaurus Isis" lived 50-30 million years ago. It had legs! Knee joints, feet and toes! It was nicknamed the "walking whale".

High-powered core drill tool marks were found used in the Giza plateau over 100 years ago. Power saw marks were also discovered cutting into granite. Even *mistakes* were found! (see photo). It's laughable that a supposed Egyptologist and Director of Ancient Egypt Research Associates (AERA)- Dr. Mark Lehner - said -

"Hey folks, these weren't lasers. These were chisels and hammers".

How can he be considered an expert when he can't even be honest? It's IMPOSSIBLE to produce a hole with chisels and hammers. I'd ask Dr. Lehner to prove it on national television with any mason he chooses.

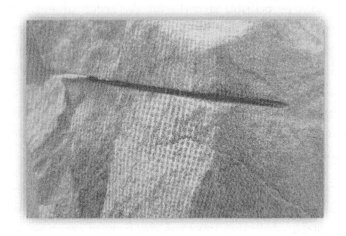

DEEP CIRCULAR SAW MARK

DUNN VS. CADMAN

Christopher Dunn is an author and engineer with over 35 years of experience with electrical and mechanical engineering. In 1969 he worked for an aerospace company as a skilled machinist and toolmaker. He served as a Project Engineer and Laser Operations Manager at Danville Metal Stamping, a Midwest aerospace manufacturer. He has written several book on Egyptology - specifically about the Giza pyramids.

John Cadman is an engineer hailing from the Pacific Northwest. He is primarily responsible for developing the theory that the Giza Pyramids served as a hydraulic pulse generator. He has built large-scale models of the Giza Pyramid's lower section, attempting to prove his theories.

Engineer	Primary Purpose	Chemicals Used	Fuels Created	Chambers Used	Evidence Left
Dunn	Electricity	Hydrochloric Acid, Zinc Ca	Hydrogen, Sound	Kings, Queens	Residue of gypsum
Cadman	Water Pump	Catalyst		Pit well shaft?	Salt, iron, gold

Author Spencer Cross discusses an interesting process that he summarizes could have taken place using the following materials found in one of the shafts leading to the Queens Chamber.

· Small bronze grapnel hook

· Piece of wood

· Stone ball

Many of you may have seen a fascinating segment on The History Channel regarding the Great Pyramid, when Rudolph Gantenbrink guided a miniature robot in 1993 up a 9" square shaft that lead towards the Queen's Chamber. In the segment, the robot was stopped by a block (or what looked like a door) made of limestone with two "*mysterious copper fittings* protruding through it". Here's Spencer Cross's quote –

"While I was watching the video of the exploration with my friend Jeff Summers, he off-handedly remarked that the fitting looked like electrodes. His observation made sense to me. To deliver an accurate measure of hydrochloric solution (See Dunn's theory in the chart) to the reaction chamber, a certain head pressure would need to be maintained. The head pressure is determined by the volume of fluid in the channel that determines the weight of the column of chemicals. The copper fittings would have served as a switch to signal the need for more chemicals. Floating on the surface of the fluid would have been another part of this switch – the cedar-like wood joined together with the bronze grapnel hook. This assembly would rise and fall with the fluid in the channel. With the channel full, the bronze prongs would have made contact with the electrodes, creating a circuit, and as the fluid in the channel dropped, the prongs would move away

from the electrodes, thereby breaking the contact and acting as a switch to signal the pumping of more chemical solution into the channel until the bronze hook again made contact with the electrodes. As the rate of supply into the reaction chamber was slight, a small opening was all that was needed to maintain the supply of chemicals".

Spencer L Cross – "The Great Pyramid: A Factory for Mono-atomic Gold" (Pg.120)

What a great analysis! To all of those who barely passed Chemistry like me, this analysis makes complete sense. Read it again….. slowly. It is marvelous. . So now I ask all of you skeptics again – if this hypothesis is true (or at least part of it), is it *remotely possible* for Jewish slaves to have built it?

CADMAN

Cadman believed that the primary function of the Great Pyramid was to act as a pump. In summary, the process has 9 steps as follows:

1. Salt is carried by cart and is lowered by pulley system into the PIT AREA.

2. Fresh water is poured down into the PIT and dissolves the salt.

3. PUMP SYSTEM is activated.

4. Valves are turned on which transport the salt water into the Kings Chamber's southern shaft.

5. Then it is released into the "wall" of the Kings Chamber, and "seeps into it".

6. The Baghdad battery is then deployed to activate electrolysis.

7. Shafts would then electrolyze (gold plated)

8. The salt would then be converted to sodium hydroxide, then to chlorine gas.

9. The chlorine gas is then diverted to the Queens Chamber and is converted to hydrochloric acid, which leaves the following:

THE KING'S CHAMBER – sodium hydroxide

THE QUEEN'S CHAMBER – hydrochloric acid

THE PIT – is initially clean, but then is used to mix the hydrochloric acid with salt water and gold to convert to hydrogen peroxide. The hydrochloric acid and hydrogen peroxide mixes with (none other than) gold to produce a *honey-type solution*, which mixes with sodium hydroxide to become a reddish solution. The colors then transform through wild swings in its pH to eventually become a WHITE CAKE. The WHITE CAKE is what the Bible refers to as "MANNA" which is to be consumed! Remember the Jews wanted to go into the land of "milk and honey"? Well, the "honey" is represented by the honey-type solution that is made in THE PIT.

THE MANNA - is essentially "mono-atomic gold". Isn't it interesting that the Jews, according to the Old Testament, survived for "forty years" on manna alone! Can this be? An edible form of gold? It seems like an awful long time – forty years – to be eating mono-atomic gold....

In summary – Dunn believed that the Pyramid's purpose was that of an *Electrical Power Plant of Hydrogen*, using both the Kings and Queens Chambers. Whereas Cadman's belief was that it existed as a *Water Pump*, or a *Vertical Pump*, using an area in the pyramid called – "The Pit". The problem with Cadman's theory is that it doesn't explain the *residue* left in the pyramid.

Spencer Cross describes it beautifully in "The Great Pyramid – a factory for mono-atomic gold" – Dunn "identifies how the use of hydrochloric acid and some corresponding chemicals were used to create hydrogen fuel, which the Egyptians or another civilization consumed at the time that it was operational". Whereas Cadman states that the "Great Pyramid was designed as a water pump which had the ability to force water vertically".

Egyptians **restored**, not built the pyramids. Christopher Dunn, a real expert Egyptologist and electrical engineer, has speculated that a hydrogen explosion took place in the Grand Gallery. Jerret Gardner explains in "What Egyptologists Don't Want You to See" –

"The hydrogen he believes was produced in the Queen's Chamber. The discontinued shafts (he believes) were used to dispense dilute hydrochloric acid and zinc chloride solutions into the chamber. When these chemicals mix they create hydrogen gas. However, there's a salt by-product. These internal pyramid mutations are definite signs of a destructive breakdown, but what of a possible meltdown? The chances are unlikely considering the only proposed power production methods are electromagnetic, solar, kinetic conversion, and hydrogen. Meltdowns are a term given solely to describe the severe overheating of a nuclear reactor core.

"However, the Ring of Fire could have caused a meltdown-like result. It's a flowing river of molten lava running below the Earth's crust. Its path can be traced all around the world and it happens to run under northern Egypt."

"Fissures appear to be the points where intense heat exited, melted the rock as it traveled and fanned out, and then later hardened into the thick black and red static rivers we see today."

There has been discovered a secret subterranean man-made tunnel complex at an area called "NC2" – or the "North Cliff 2". It is located in the northwest edge of the Giza Plateau. Inside there are secret sub-floor excavations that have never been revealed to the public.

The word "Pyramid" in Arabic means "ultimate age or size"

The pyramids were built by one of Enki's sons, Thoth, right after the great flood of 11,500BC. Gibil, another of Enki's sons, "installed pulsating crystals and a 32 ft. capstone (missing today - you can see where the top has been replaced) made of electrum to reflect a beam of light into space for the incoming spacecrafts" (King). The Sphinx's face is originally that of Thoth, but was replaced instead by Marduk, Enki's villainous son to that of his son named Ansar. It is the face of Ansar that exists on the Sphinx today.

Evidence for the date of the pyramids being built in 11,500BC includes hundreds of thousands of marine fossils and seashells embedded in the walls of the pyramids and the surrounding monuments. Joseph Jochmans explains that geologists *"are hard pressed to explain why there existed a fourteen-foot layer of salt sediment around the base of the pyramid, a layer which also contained many seashells, and the fossil of a sea cow, all of which are dated by radiocarbon methods to 11,600 B.P.(Before Present), plus or minus 300 years*". Haze - "Aliens in Ancient Egypt" pg. 18

PYRAMIDS AS MACHINES

Plate tectonics demonstrates that the shifting of the earth's plates causes seismic energy. Charged molten metal circulating in the core produces an electric current that generates a magnetic field. This magnetic field runs along geometric grid lines that are capable of being harnessed. Imagine if geniuses like Nikola Tesla had known this and were able to tap into this energy and use it to direct wireless energy!

So, are the pyramids acting as receivers of the energy already emitting from the earth's grid lines?

Dunn puts it beautifully -

"The Great Pyramid was a geomechanical power plant that responded sympathetically with the Earth's vibrations and converted that energy into electricity. They used the electricity to power their civilization, which included machine tools with which they shaped hard, igneous rock." - (Dunn - The Giza Power Plant)

The earth also emits a pulse, or a sound, generated by waves of mechanical, thermal, electrical, magnetic, nuclear and chemical reactions. Sound is also generated by piezoelectrical materials such as quartz, which is found in Giza. It is interesting to note that based upon a study of the King's chamber headed up by Tom Danley, a NASA consultant and acoustical engineer, the note that resonates is an F sharp.

There exist interesting and mysterious large crystal bowls embedded in the sand next to the pyramids. What is their purpose? Quartz crystals and a quartz floor make up their basin. We know from my previous book - *"Adam = Alien"* that quartz crystal can carry electrical signals. We know that the chambers in the pyramids are harmonically tuned to a specific frequency or musical tone. We also know from *"Adam = Alien"* that it is possible to levitate heavy objects with specific sounds and frequencies. Did the bowls utilize its quartz content and shape to function almost as machine? Were they used in conduction with the Earth's "ley lines" which are energy currents that invisibly surround the Earth? Did the quartz crystal channel this energy, creating natural energy fields?

Knowing that the Giza pyramids are made of a type of limestone called dolomite, which has a high magnesium content, could it be that we have a highly electrically conductive core wrapped in a very effective insulator? With the help of the granite passageways, the pyramid could become radioactive and ionize or electrify the air.

Today, quartz crystals are used in AM/FM radios, CD's, DVD's and computer processors. The pyramids may have acted as machines that emitted grand-scale signals - that may be capable of reaching into space.

It is thought that the transference crystals that has been written about in Sumerian tablets, working inside of the pyramids, function as transmitters of signals sent to Nibiru from Earth, and gold helped improve the transmission.

ALIEN IN ACTUAL EGYPTIAN HIEROGLYPH

INSIDE THE GREAT PYRAMID - CORBELLED
CEILING EHANCED SOUND TRANSMISSION

PYRAMIDS AS REPRESENTATION OF HUMAN PERFECTION

In the book – "The Temple in Man", author R.A. Schwaller de Lubicz describes the Temple of Luxor as a representation of the human body. From its physical layout, to its spiritual properties. Schwaller de Lubicz spent 15 years there and mapped out every square inch of the pyramid and the surrounding monoliths. He suggests that *numbers* did not simply designate quantities but rather *"were concrete definitions of energetic formative principles of nature."* The Egyptians called these energetic principles *Neters*, a word which is conventionally rendered as "gods". He introduces the notion of the "principle of the crossing" to describe mathematical multiplication. That the symbol we use today in multiplication (X). For example – 2 X 1 = 2. The crossing of two 1's is equal to the number 2. So "1" refers to a neter, and two neters, represents the crossing of 1's – hence the X symbol. I find it interesting that the letter "X" is, literally, the crossing of two "I"'s. One I is slanted across the other I.

The unity of male and female in procreation can be symbolized by the "X", so as to "multiply" and create more humans. If 1 neter represents a god, and the X represents two neters combined, then the number 2 represents the procreation of the 1. This is combined because of organic creative energy. Schwaller de Lubicz suggests that the architecture of the Temple of Luxor provides indisputable evidence of a divine symbolic directive.

The pattern of the pyramid is irregular – de Lubicz found 1001 irregularities. Just as Washington D.C. has patterns in its architecture to suggest isosceles triangles drawn between the White House, The Capital and Congress, the Temple of Luxor shows isosceles triangles between a human bodies heart, lungs and brain. Each axis shows a theme connecting the organs of the human body. There is spatial and even ideological orientation in its layout. Schwaller de Lubicz writes –

"The Outline of a human skeleton – traced according to anthropometrical methods and very carefully constructed bone by bone – was superimposed on the general plan of the temple."

The number "3" is also represented as the "trinity", or the combination of three "1"'s. It's interesting that in ancient China the number one was considered to have a value of three. There seems to be evidence that the same applies to the Egyptian value of one. The Egyptian "trinity" was a symbol not of 3 but of 1. Man composed of 3 beings – the sexual being, the corporeal being and the spiritual being. Even the pyramid shape has 3 sides but symbolizes 1.

Why do the Egyptians draw figures sideways on stone? Why don't they draw figures straight-on as they appear in real life? Why are the legs and arms separated one in front of another, right hand in front of left, and right foot in front of the left foot? The answer is that the Egyptians were suggesting movement or action. The right hand gives, while the left hand receives. The legs are separated indicating movement (as opposed to mummies which have legs together, indicating death or inertia). The significance was life versus death.

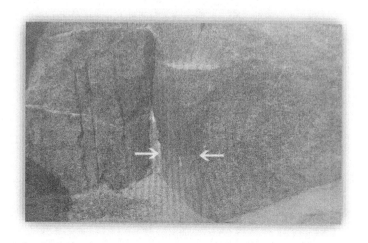

EVIDENCE OF A SMOOTH CUT ON GRANITE
WEIGHING SEVERAL TONS. THIS CUT IS TOO
SMOOTH FOR COPPER TOOLS TO CREATE

EVIDENCE OF PERFECT WAVE CUT THROUGH
9FT OF HEAVY SANDSTONE. ONLY A PERFECT
LASER COULD HAVE CREATED THIS.

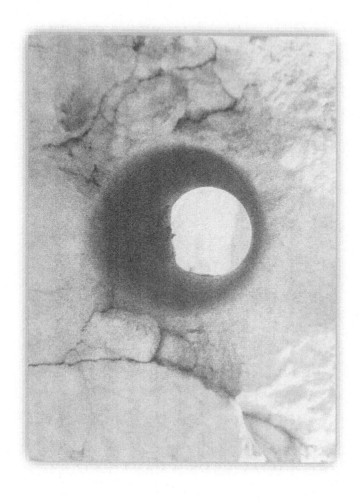

EVIDENCE OF PERFECT CIRCULAR DRILL
HOLE. IMPOSSIBLE TO CREATE IN 1500 BC

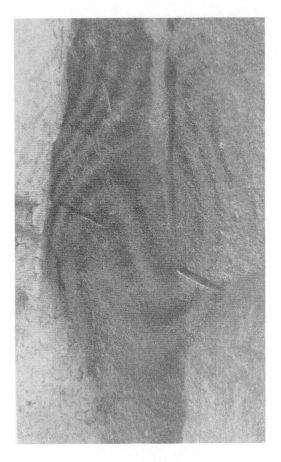

TWO CUT MARKS OF A MASSIVE CIRCULAR
SAW

SOUND ♫

Dunn writes in *The Giza Power Plant* -

"The granite out of which the King's Chamber is constructed is an igneous rock containing silicon-quartz crystals. This particular granite, which was brought from the Aswan quarries, contains fifty-five percent or more quartz crystal" (Dunn - The Giza Power Plant)

This crystal permits a piezoelectric signal to carry through it, meaning that energy can be transmitted across its entire body/length. The granite readies itself to convert vibrations from the earth and convert them into electricity. Dunn continues -

"(The ancients) had determined that they needed to tap into the vibrations of the Earth over a larger area inside the pyramid to deliver that energy to the power center - the King's Chamber - thereby substantially increasing the amplitude of the oscillations of the granite...The Great Pyramid can be seen as a huge musical instrument with each element designed to enhance the performance of the other....The Grand Gallery, which is considered to be an architectural masterpiece, is an enclosed space in which resonators were installed in the slots along the edge that runs the length of the gallery...Thus, with the input of sound and the maximization of resonance, the entire granite complex in effect, became a vibrating mass of energy." - (Dunn - The Giza Power Plant)

He backs this up by stating that visitors and researchers alike have noted peculiar sounds reverberating inside of the pyramid. When Napoleon Bonaparte travelled from France to Egypt, his men shot pistols at the top of the Grand Gallery, and "noted that the explosion reverberated into the distance like rolling thunder." (Dunn - page 161). Dunn actually struck the coffer inside the King's Chamber and heard a deep, bell-like sound. Using a tuning fork, he was able to determine that the note was an "A", vibrating at 438 cycles per second. This note actually carried from the Grand Gallery, through the passageway, and reverberated inside the King's Chamber.

He goes on to describe the use of 30,000 stone bowls found in chambers underneath the Step Pyramid, some with handles, some without, and one in particular that has a *stone horn* attached used as what he describes as a Helmholz resonator. The Helmholtz resonator is a modern apparatus that responds to vibrations and maximizes the transfer of energy from the source of the vibrations. Additionally, an acoustical engineer confirmed that the Antechamber acts as an acoustical filter of this sound. Can it be that the Great Pyramid acted as a resonance generating power plant? Remarkable! Can this be coincidence?

Dunn provides conclusive evidence that the Great Pyramid generates hydrogen. He has found residue of hydrated zinc chloride and diluted hydrochloric acid in the Queen's Chamber, used chemically to produce hydrogen. Evidence of salt encrusted on the walls of the Queen's Chamber is proof of this process. The Chamber is also designed to take in fluid rather than air.

He summarizes beautifully as follows -

"...Support my premise that the Great Pyramid was a power plant and the King's Chamber its power center. Facilitated by the element that fuels our sun (hydrogen) and uniting the energy of the universe with that of the Earth, the ancient Egyptians converted vibrational energy into microwave energy. For the power plant to function, the designers and operators had to induce vibration in the Great Pyramid that was in tune with the harmonic resonant vibrations of the Earth. Once the pyramid was vibrating in tune with the Earth's pulse it became a coupled oscillator and could sustain the transfer of energy from the Earth with little or no feedback." (Dunn - The Giza Power Plant)

EGYPTOLOGISTS
FRAUD

Egyptologists will tell you that the pharaohs were buried inside of the pyramids. According to Sitchin, there is only one instance in which a mummified pharaoh is buried inside of a tomb. The instance was inside the "small pyramid" at Giza. Sitchin explains as follows:

"In July 1837 an Englishman by the name of Howard Vyse, who was excavating in the area, reported that he had found near a stone sarcophagus inside this pyramid fragments - the cover of a mummy case with a royal inscription on it, together with the part of a skeleton of the king's name. The name was spelled out MEN-KA-RA....it was proved to have been an archaeological fraud. Scholars at the time already had some doubts about the age of the mummy case due to its style. And when a few decades ago radio-carbon dating was developed, it was established without doubt that the mummy-case cover belonged not to the fourth but to the twenty-fifth dynasty, not to 2600 BC but to 700 BC., and not even from the pre-Christian times, but from the first centuries of the Christian era. Someone, in other words, took a piece of wooden coffin found elsewhere, and a skeleton from a common grave, and put them in a pile of rubble inside the Small Pyramid and announced: Look what I found!"

Sitchin goes on to say…

"As I saw it, the Anunnaki and not the pharaohs had built the Giza pyramids and carved out the Sphinx, and they did it not circa 2600 BC, but circa 9000 BC. But that is not what the Egyptologists were saying."

Similarly, we find fraud evidenced in the Great Pyramid. There are no inscriptions or markings in the Great Pyramid, however, Howard Vyse once again enters the scene. Three inscriptions in red paint were found spelling the name "Khufu" or the pharaoh Cheops. This led us to believe that Cheops was the builder of the Great Pyramid. This too was a hoax! Sitchin was able to find a copy of the cloth copy of the inscription at the British Museum in London and examined it. He found that Howard Vyse's assistant named Mr. Hill had mistakenly painted the name "Raufu" instead of "Khufu"!

There is no evidence to support that the Pyramids were used as tombs or that they were even built by the Egyptians. Proof of the existence of UFO's in different cultures' texts and writings:

1.   Egyptian UFO's – winged discs – Eye of Horus
2.   Hebrew – Ezekiel, Enoch and Elijah – trips aboard flying chariots
3.   Hindu – flying vimanas
4.   Persia – flying sun disc
5.   Ezekiel and Moses - burning bush or a wheel-within-a-wheel

Olsen discusses the impossibility of humans creating the pyramids alone -

"...The four air shafts of the Great Pyramid point precisely to the key stars of Orion, Sirius, Thuban, and Polaris as seen from the time frame of the Osirius Empire. The alignment of the Giza pyramids on the ground matches perfectly with the alignment of the constellation of Orion as seen in the sky from Giza, relative to the Nile River, as the Earthly representation of the Milky Way galaxy in the night sky. The placement on the ground of these structures has geodetic or astronomical significance relative to the stars in our Galactic region. The configuration of the three main pyramids on the Giza Plateau was intended to create a mirror image on Earth of our solar system with certain key constellations." - Olsen - Past Esoteric - page 170.

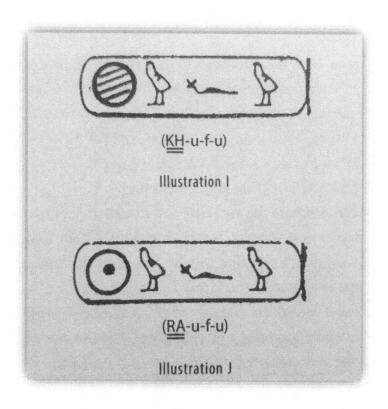

(KH-u-f-u)

Illustration I

(RA-u-f-u)

Illustration J

CHEOPS HOAX- THIS WAS WHAT MR. HILL
PAINTED IN THE GREAT PYRAMID-RAUFU
INSTEAD OF KHUFU

EVIDENCE OF BORING HOLE CYLINDRICAL CUT

CHAPTER 3

THE FIRST

AT FIRST, ENKI CREATED FOUR men and four women, all different and unique. These four represented the first Caucasian, Mongoloid, Australoid and Negroid. The Caucasoid and Mongoloids did all the laborious agriculture work, and the Australoid and Negroids did all the mining. They worshiped their gods and were essentially slaves.

The story of Adam and Eve has been historically twisted. However, elements of the story are preserved in the Sumerian tablets and have been written to support the needs of the culture that ruled at the time. Based on the long-standing feud between Enki and Enlil, certain elements of the story show the inter-relationships between Enki, Enlil and the proto-humans. Based on the supposition that Enki did in fact create the first humans and cared for them as his own, there are many sub-plots that developed. The proto-humans were slaves used for agriculture and mining, so Enlil needed their labor. Enki was interested in educating and developing them into sentient beings. It is written that when Eve told Enki that Enlil "would destroy any primitive worker whoever defied their commands" (tablets). At this, Enki declared, "I am your creator, and Anu and Enlil will not harm you or Adam". Eve, now believing that Enki is "god" asks him about the sacred "knowledge" - which I (the author) believe means both that at a higher level the Anunnaki are immortal and have the knowledge of immortality, and at a lesser level, sexual procurement - has intercourse with Enki. Afterwards, Eve knowing and enjoying the sexual act, has intercourse with Adam. Enki represents the snake in the Bible and is depicted throughout the world on columns, pyramids, buildings and tablets. The "apple" represented sex, and when Adam and Eve "eat" the apple, it is a representation of intercourse. "Fruit" in the Bible also carries the meaning of intercourse.

Upon learning that Enki visited Eve, Enlil banishes them from the Garden of Edin. Enki covers their bodies with white cloth and sends them on their way. Eve then gives birth to Cain who the author believes is the actual son of Enki, then to Abel who the author believes is the actual son of Adam. Cain carries the actual bloodline of the gods because he is half-divine. When Enlil asks Cain and Abel to prepare a feast for him, Cain offers produce, while Abel offers a sheep. Enlil rejects Cain's offer and accepts Abel's, infuriating Cain who then kills Abel, after Abel's sheep destroys Cain's crops. Upon returning from his feast, Enlil asks Cain, "Where is your brother?", Cain replies. "Am I my brother's keeper?". All the while Enlil had already known that Cain killed Abel.

MASSIVE PALACE OF UR - POSSIBLY BUILT
BY/FOR GILGAMESH - STILL STANDS TODAY

THE SUMERIANS

 etween 1889 and 1900, over 30,000 clay tablets were unearthed in Nippur (today's Iraq), by the University of Pennsylvania, dating back to 1750 BC. These clay tablets with cuneiform writing is the oldest written literature on earth. This point is indisputable. This literature was translated and used by the Hittites, then Hebrews, then the Greeks. Yet the Bible, written almost two thousand years later, never speaks about this ancient Sumerian civilization or the tablets. Our History textbooks barely mention this civilization - especially considering the fact that archaeological findings at sites such as Jericho and more recently, Gobekli Tepe in Turkey, which has been dated to 11,600 years ago! Why?

The word *Sumerian* is derived from the region of southern Mesopotamia established in 4000BC called Sumer (pronounced *Shumer*). One of the foremost cities in Sumer was called Uruk (modern day Iraq), settled in 3800BC. Uruk was the first true city on Earth.

The tablets describe the great flood of Noah in Shuruppak - is this the source document for the Bible? The tablets describe the epic tale of the greatest Sumerian hero - Gilgamesh, the forerunner of Hercules in Greece, yet few know the name Gilgamesh. Did the

Bible borrow these stories from the Sumerian tablets? Was the Bible modified from, edited from, and originated from these Sumerian tablets? Why have so few people heard about these invaluable tablets? Why have they remained largely unknown, and why have they not been made available to both scholars and everyday classroom students? To quote one of the most distinguished scholars of Sumerian literature - Dr. Samuel Noah Kramer of the University of Pennsylvania -

"It soon became evident that some of the Old Testament material was mythological in character because it presented clear parallels and resembles the myths recovered from Egyptian and Babylonian sources" **(Kramer - Sumerian Mythology)**

One of the first translations by Kramer of Plate IX (tablet 13877), states:

"At one time heaven and earth were united. Some of the *gods* existed before the separation of heaven and earth"

Notice the word *gods* (plural), not god (singular). It goes on to describe the gods as *Anu* - carrier of heaven, *Enlil* - the air god, *Enki* - the water god, *Utu* - the warrior god, and the female *Nammu* - the mother god. Five gods described! How can this be? We are only told to believe that only one god exists.

"In the beginning *Elohim* created the heavens and the earth".

Elohim is a plural word. The singular would be Eloah. Why use Elohim? Eloah as a singular tense is used 250 times. It is deliberate. Elohim denotes **many** gods. Elohim is used 2,500 times! (Rapha - Aliens, Angels and Gods).

"Behold, the man has become like one of *us*" (Genesis 3:22).

Again, we see reference to the plural. When I read this, I think of the Greek pantheon of gods, sitting in a forum, deliberating on the future and welfare of their humans.

"Do not worship *any other* god, for the Lord (Yahweh), whose name is jealous, is a jealous god" (Exodus 34:14)

Any other denotes the plural again. Notice that Yahweh does not say "Do not worship any other idols" here - he says "god". Yahweh himself is admitting to the existence of other gods! How can this be? How does this play into the monotheistic identity of God?

"Canst thou bind the sweet influences of Pleiades, or loose the bands of Orion? Canst thou bring forth Mazzaroth in his season? Or cast thou guide Arcturus with his sons?" (Job 38:31-32)

Pleiades, Orion and Arcturus are described in Job. These star systems were not discovered until the 19th century! These star systems are only new discoveries. How can they be identified in 2000 BC? The only rational reason being that the scribes knew about these star system from the gods who originated themselves from here. The Anunnaki have been described as originally being from Orion and eventually migrating to Nibiru their current planet.

CUNEIFORM TABLET 1

CUNEIFORM TABLET 2

SUMERIAN KNOWLEDGE

"Sumerian writing is the oldest sophisticated form of writing in existence, having first appeared in about 3400 BC, but it is neither crude nor primitive, and there is no region on Earth which identifies any scribal concept that might have been its forerunner. It appeared in a complete and composite form, as if from another world, in the style known as cuneiform (wedge-shaped). This was a series of angular phonetic symbols (cuneates) ostensibly abbreviated from the pictographs of the Sumerian Temple priests." (Gardner - pg. 53)

The following facts will demonstrate that the Sumerians were so far advanced than the generations of homo sapiens before them, that they could not have just adapted or improved upon the generation proceeding them. They were so organized, civilized, and intelligent that they produced more "firsts" than all groups combined. But their contributions to mathematics and astronomy are so advanced that there is no way they could have been solely human. Their advances had to have been taught to them by the Anunnaki.

I believe that the following advances were initiated and executed by the Anunnaki using Sumerians as slave labor. They could not have been solely human-

- ❖ the creation of cities
- ❖ the creation of the cuneiform system of writing
- ❖ the creation of organized commerce
- ❖ the creation of canals and aqueducts
- ❖ the creation of drainage and sewage systems
- ❖ the creation of palaces
- ❖ open-water navigation
- ❖ the creation of weights and measures
- ❖ the creation of wheeled carts, wagons and chariots
- ❖ the creation of written laws, courts and judges

❖ the creation of business contracts
❖ the creation of marriage and divorce
❖ the creation of musical instruments
❖ the creation of poetry, song and dance

Their mathematics was based on the sexagesimal (60) system. I believe firmly that it is based on this number because it took Nibiru 3,600 (multiple of 60) years to pass around the Earth. But here is where I believe that the Anunnaki had a clear imprint on human knowledge, specifically relating to astronomy.

According to the Sumerian tablets, the Sumerians believed that:

1. Precession - The exact year in which Earth's polar axis points to the North Star - of 25,920 (again an exact multiple of 60) years, which they based their calendars on.
2. Shape of the Earth as round with an equator and two poles.
3. Distances between stars (an example is taken from Sumerian text #AO.6478 which lists the 26 stars visible along the line of the Tropic of Cancer)
4. The description of Uranus and Neptune as "watery twins" (how could they know that these planets contained water?), with a "blue-green" color (which cannot be seen with the naked eye). Author's note - (NASA's Voyager satellite voyage

of 1986 showed exactly this! Uranus was watery and was colored blue-green!)

Lloyd Pye summarizes as follows:

"By now, hopefully, you should be willing to at least tentatively accept that: (1) Sumerians were telling the truth in their cuneiform inscriptions about Earth's distant past; (2.) Nibiru was and is a real planet in our solar system; (3) its inhabitants were and are the Anunnaki; (4) the Anunnaki were and are a race of highly advanced beings; and (5) they lived and worked on Earth as the dominant culture from 430,000 years ago until 2000 BC., when a nuclear war broke out among them, and from which they never recovered, causing them to leave Earth en masse in 200 BC. "
('Everything You Know is Wrong" - Pye page 246)

It is also interesting that the Sumerians referred to their gods as "the lofty ones" - an interesting description of their hierarchical position within the Sumerian society and that they were *aloft or* flew in the air!

CIRCULAR TABLET WITH CUNEIFORM
WRITING

RECTANGULAR TABLET

CHAPTER 4

THE NIBIRUANS

𝕀𝕀𝕀 𐎟 𐎟 𐎟 𐎟 𐎟 𐎟 𐎟 𐎟 𐎟 𐎟 𐎟 𐎟

I HAVE WRITTEN EXTENSIVELY ABOUT the Anunnaki in my previous book - "Adam=Alien", their contact with Earth, their need for gold and their seeding the human race. They are an extraordinary species of humans originating from the planet Nibiru in our very own solar system, not far from Earth. They are written about extensively in the Old Testament - referred to as the *Anakim* in Hebrew. It is believed that Nibiru suffered a catastrophic shockwave in its original orbit of Canis Major, when a red giant star called Sirius B or Osiris (yes, same as in Egyptian lore) imploded and ejected Nibiru out of Canis Major. It filled Nibiru's orbit with a super-abundance of radioactive heavy metals. The ejection sent it right towards Earth's solar system, sending it to revolve in a *clockwise* fashion elliptically, which was against the current circular, counter-clockwise orbits of our 9 planetary bodies. Because of this abundance of radioactive heavy metals, Nibiru's atmosphere began to wither, thus threatening the Anunnaki with extinction. They had to act quickly. They initially created a base on Mars, which was the closest and most habitable planet nearest to them. Based on Nibiru's new elliptical orbit which passes our planets every 3,600 years, Nibiru has been called the "Planet of the Crossing". The ankh symbol with its cross and circular top, signifies the planet of the cross(ing), and some writers believe that the cross of Christianity also represents Nibiru.

The Anunnaki built 12 major cities on Earth (12 being multiple of the number 6 which was the standard sexagesimal number of their mathematics). These 12 cities resided predominantly in current day Iraq - Kish, Uruk, Ur, Sippur, Akshak, Larak, Nippur, Adab, Umma, Lagash, Bad-Tabira, and Larsa. They built huge temples (similar to the ziggurat) in the center of each of these cities which housed the Gods.

The Anunnaki also formed an Elite Aristocracy known as the Council of 12. These 12 gods were Anu, Enlil, Enki, Nannar, Utu, Ishkur, Antu, Ninlil, Ningal, Inanna and Ninhursag. This council concluded that a heat shield made up of light precious metals (including gold) was needed to save their planet. They had initially mined these light precious metals on Mars in the Cydonia region - where the "face" and the "city" have been photographed by NASA. They used Mars as a storage facility after mining began on Earth.

They are war-like beings with narcissistic tendencies, extremely driven to gain power, control and influence over territory and governance. According to author Alex Collier, the origin of the Anunnaki is as follows:

"War-like factions from Sirius and Orion decided to broker a peace accord by inter-marrying. A male king from Sirius B married a queen from Orion, their offspring became a race named "Nibiru" or "divided amongst two". They flourished on their own planet which thenceforth became named Nibiru."

1. Gold is reason for Anunnaki coming to Earth, hence the high value put onto it by all humans.
2. Anunnaki used the Earth's free energy as its resource to mine the gold
3. The Ark of the Covenant levitated and was pushed by 4 men
4. White Powder of Gold – the ingredient of "Manna", is a powerful substance.
5. The Golden Calf was not melted, it was grinded down into white gold, which was dissolved in water and drunk by the Israelites.
6. A gold mine in South Africa revealed the bones of a 50,000-year-old homo sapien child
7. Stone axes have been discovered dating back to 200,000 – 400,000 years old which fits into my last book *"Adam = Aliens"* claim that humans were genetically engineered 440,000 years ago.

According to Michael Tellinger in his book – "African Temples of the Anunnaki", regarding the power of white monoatomic gold – "*This is the information that the Massachusetts Institute of Technology (MIT) has been very secretive about...the healing properties of the powder are most mysterious and is most likely what Moses was doing in the desert...the presence of the white light seems to repair all genetic defects in our DNA, and it heals human cells from any disease that may exhibit. This is most likely what Royal Raymond Rife discovered in 1931, when he reportedly found the 'cure for all disease'"* – (pg. 114)

Tellinger found tens of thousands of large circular, rose-like shaped, formations in South Africa. He compared the shape to that of a "Resonant cavity *magnetron* (High-powered high frequency), oscillator. How interesting! He compares the shape and is very convincing! He states that the long lines of rocks leading out of the rose-shaped cavity are not roads, but rather "energy channels – connection devices" (pg118). He continues (pg. 121)

"The magnetron – which is used for generating sound with vibrational frequency energy."

Energy in many modern appliances like microwaves – is virtually a copy of many of the flower-shaped stone structures. The magnetron's relative, the klystron, also has myriad applications, such as radar…The central rod or stone, is made to vibrate at a specific frequency that is amplified in the adjacent resonant chambers and then channeled out via the connectors that conduct the vibrational energy to another destination, where it is used in many possible ways. The original vibration in the center of the magnetron, or circle, can be generated by sound. The frequency or pitch of the sound will create the specific energy required to perform various tasks. The tasks can vary from:

1. Magnetism
2. Drilling
3. Levitation

John Worrell Keely used a magnetron in 1888 to perform these functions, so it has been used in the USA for 120 years in appliances such as our microwave and even in radar!

EVIDENCE

1.   IRI REIS MAP – BASED ON its mapping of Antarctica, the only way it could have been determined was from the sky prior to 4,000BC, prior to Queen Maud Land being covered with ice. Shows Andes Mountains on western side of South America which were unknown in the 1500's and the Amazon River rising in the Andes and flowing eastwards.

2. Nazca Lines – it appears that the famous "Spider Figure" is a terrestrial diagram of the constellation of Orion. The condor figure is 400 feet long, the hummingbird is 165 feet long and the spider is 150 feet long. These were no mere doodling's from ancient Indian tribesmen!

3. Viracocha- means "Foam of the Sea"- describes Enki as "a bearded man of tall stature clothed in a white robe which came down to his feet and which he wore belted at the waist", a white man with blue eyes, and –

"A bearded man of medium height dressed in a rather long cloak…He was past his prime, with grey hair, and lean. He walked with a staff and addressed the natives with love, calling them his sons and daughters. As he traversed all the land he worked miracles. He healed the sick by touch. He spoke every tongue even better than the natives." – "Fingerprints of the Gods – Hancock"

4. Quetzalcoatl – means "Feathered Serpent" – is this Viracocha/Enki in the Andes? Introduced writing, the calendar, astronomy, medicine, masonry, architecture, math, metallurgy and introduced corn – which had been *foreign to Earth* – and *brought to Earth*

5. Olmec head figures depict clearly negroid features. But there were no Africans in the Americas prior to 2000 years ago. How did they get there? Olmecs were also the inventor of the calendar – not the Maya. They were much older than the Maya and invented the starting date of August 13, 3114 BC as the start of civilized life ending in 2012 AD.

Constant themes found in Mexico and Central/South America are:

1. Bearded men
2. Crosses
3. Serpents

My theory is that they crop up repeatedly to remind us that:

1. We are the children of the Anunnaki

2. Our mother planet is Nibiru (the planet of the Crossing in the Sumerian tablets)

3. We were formed through their DNA (entwined serpent)

Pyramid Connection – Hancock brilliantly suggests a mirroring of pyramids in Giza and Mexico- *"Just as at Giza, three principle pyramids had been built at Teotihuacan: The Pyramid/Temple of Quetzalcoatl, the Pyramid of the Sun and the Pyramid of the Moon. Just as at Giza, the site plan was not symmetrical, as one might have expected, but involved two structures in different alignment with each other while the third appeared to have been deliberately offset to one side....could this be coincidence?"* – Hancock – Fingerprints of the Gods – pg. 169

VIRACOCHA

QUETZALCOATL

NIBIRU

I ALWAYS ASKED MYSELF HOW THE Anunnaki, with a similar physical makeup to humans, could survive on planet Nibiru which was so far from Pluto in an atmosphere that was extremely cold? The answer - their planet had an *internal* source of heat. Even though they were so far from the sun's rays, they could exist comfortably. Nibiru passes the Earth every 3,600 years. It is only 51 light years from Earth, or 250 trillion miles away….not to distant in relation to our galaxy.

The symbol of Nibiru is a cross (+), which symbolizes Nibiru crossing the Earth. Nibiru holds the key to the major Earth extinctions, floods, climate changes etc., due to its passing of the Earth. According to Alford-

"According to the Enuma Elish, Nibiru was forever destined to return to the place of the celestial battle, where it had crossed the path of Tiamat (Earth's original name) - it was for this reason that it became known as the 'Planet of the Crossing'.…the cross, sacred to Buddhism as well as to Christendom, thus owes its origin to the celestial event which created the Earth and the Heavens." (Alan Alford - pg. 227).

Nibiru's existence is well known to astronomers. It is called "Planet X". It has been acknowledged in publications such as The New York Times, the Washington Post, National Geographic, and numerous other scientific journals as a planet capable of sustaining life.

So, I ask you, dear reader, if Planet X is real, exists in our galaxy, can sustain life, can be seen by the Hubble Telescope (and other telescopes).....isn't it possible, if not probable, that the content in the Sumerian tablets is true? Why would the Sumerians, almost 10,000 years ago make up a story of nine-foot-tall gods arriving in spaceships and teaching them moral conduct and how to improve their existence...? They wouldn't...

It's also interesting that the Enuma Elish contains seven tablets. It was written over a thousand years before the Old Testament. I believe that the verse, "God created the Earth in seven days", refers to each of the seven tablets of the Enuma Elish.

Nibiru has made entry into our direct Solar System in the following years-

> 11,000 BC
> 7,400 BC
> 3,800 BC
> 200 BC

It's next entry will be in 3,400 AD! Well past our lifetimes! But its entry will cause a huge natural disaster on Earth because of its entry between Mars and Jupiter, and it will disrupt the counterclockwise movement of the planets by entering clockwise and will rock the gravitational homeostasis that currently exists. I am concerned about how disastrous this will be for the Earth in 3400.

Evidence suggests that the Anunnaki gods withdrew from Earth in 200 BC, since there have been no obvious signs of any physical presence now here on Earth.

Nibiru's entry into our solar system is the reason for our Earth's wobble.

It's interesting that the age difference between Nibiru and our Solar System is approximately 500 million years - exactly the amount of time that it took for Earth to acquire unicellular life forms…..coincidence?   Earth is also missing a huge portion of its crust. All other planets and moons studied contain approximately 10% outer crust, whereas the Earth contains only 1%   Can this huge discrepancy lie in the fact that the Earth was indeed smashed into by Nibiru as described in the Enuma Elish? Is this the reason that the Earth contains such deep gauges (ocean trenches)? Does Nibiru also carry such evidence from the collision?   Did Nibiru have oceans which transferred during the collision with Earth?

Modern scientists today refer to Nibiru as "Planet X". In countless newspaper articles on astronomy (including newspapers such as the New York Times and the Wall Street Journal), a planet in our solar system that is larger than Earth, and can be seen clearly by telescope, exists relatively close to Earth. These scientists know that Planet X is there based on the slight wobbles in the celestial bodies that are near to it. They also know that that planet belongs to our Solar System, and that it is indeed part of our 9 planet (actually 10 planet) family. It is referred to as Planet X because it is the *tenth* planet in our Solar System, and that it was *indeed* the Planet of the Crossing!

Some interesting data that we have learned about Nibiru is the following:

- ❖ it is approximately 40,000 miles wide (compared to Earth's 26,000)
- ❖ it generates heat from within itself
- ❖ its atmosphere is similar to Earth's
- ❖ the Anunnaki are mostly blondes, with some redheads, and have pale skin
- ❖ the Anunnaki on Earth kept their heads covered and protected by helmets
- ❖ it crosses between Mars and Jupiter once every 3600 years and radiates its own heat

NIBIRU IS HEAVEN. It all makes perfect sense. When our parents tell us that "after we die " we all go up "to heaven", now we really know that it is, in fact, a real place. A real place of our true origins…

Nibiru is trapped in a 3,600-year retrograde elliptical orbit around our sun. In 1986, the IRAS Naval Observatory telescope discovered a brown dwarf that it named "Planet X". The Vatican's own telescope called – L.U.C.I.F.E.R (interesting name, no?), corroborated the find, and confirmed its existence. It is real.

The founder of gold on Earth was originally Alalu – the disgraced Anunnaki who lost to Anu in a "fight" for the throne. Alalu had escaped Nibiru, only to discover that Earth had the gold that (if dispersed in ionized particles) could save Nibiru's failing atmosphere.....

The mining operation was moved from the Persian Gulf to the vicinity of the Zambezi River in South Africa...

Lloyd Pye summarizes as follows:

Hopefully, you should be willing to at least tentatively accept that: (1) Sumerians were telling the truth in their cuneiform inscriptions about Earth's distant past; (2.) Nibiru was and is a real planet in our solar system; (3) its inhabitants were and are the Anunnaki; (4) there Anunnaki were and are a race of highly advanced beings; and (5) they lived and worked on Earth as the dominant culture from 430,000 years ago until 2000 BC., when a nuclear war broke out among them, and from which they never recovered, causing them to leave Earth en masse at 200 BC. " ('Everything You Know is Wrong" - Pye page 246)

ACTUAL PHOTOS OF PLANET X - PLANET
NIBIRU, TAKEN FROM EARTH

Anunnaki

N.UNNA.KI – MEANS "THOSE FROM HEAVEN TO EARTH CAME"

1. The planet Nibiru has been confirmed by IRAS Naval Observatory in 1986 as a "brown dwarf planet"

2. Enki worked in South Africa near the Zambezi River

3. Anu came to Earth between 445,000 - 360,000 years ago

4. Ninhursag (Enki's half-sister) established her medical center 415,000 years ago in Shurrupak

5. The Anunnaki aged significantly on Earth due to solar radiation

6. The Watchers were the gold mine workers

7. Eridu – "Earth Station One" – was photographed by the University of Chicago in 1973 in the exact location mentioned in the Sumerian King's List

8. Sippar – in this city 400 chronologically ordered clay tablets were found describing Anunnaki history

9. Shimti – Enki's Genetics lab's location has also been corroborated back to the origins of mitochondrial DNA

10. "If he gives me clay, then I will do it, Nintu shall mix clay, with his flesh and blood, then a God and a Man, will be mixed together in clay" – (Bible reference to man made from dirt and clay)

11. Junk DNA may actually be the "turned off" balance of Anunnaki DNA that we can't figure out how to tap into

12. Uruk – where Iraq gets its name

13. Stargate to Nibiru – could it be that the invasion of Iraq and Saddam Hussein had to do with finding this Stargate?

14. Shems – rocket ships

15. Baalbek, Lebanon was the site of the Anunnaki Space facility

16. After the nuclear onslaught of the Sinai Peninsula, the Anunnaki escaped the "evil wind" to Greece creating the Greek Pantheon Council of 12- including Poseidon and the other gods

17. Enlil = Zeus, Enki = Ptah and Poseidon

18. Enki made humans with foreskins

19. Eden = Basara (on the Persian Gulf)

20. Bagdad = Babylon

21. Enlil and Marduk are enemies because the king and queen (Anu and Alalu) pledged Marduk to rule Nibiru, not Enlil

22. Enki landed on Mars and drank water from a lake
23. Enki built houses near Basara = Basra
24. The 2 secrets that the Anunnaki kept from the Adam and Eve were procreation and immortality. Eating from the tree enlightened them solely of procreation. Enki excluded the longevity gene when he made Adamu
25. After Enlil banished Adam and Eve from the Garden, Enki started the "Brotherhood of the Snake" to share secrets of advanced technology
26. Gold was transported from South African gold mines, into submersibles, to Bad-Tibira in Sumer to refine, smelt and form into ingots for transshipment to Mars
27. "The Anunnaki and their laborers sent capacitated energy generated by sound along the roads or connecting devices and moved goods and water with 'a levitation device that tapped into the magnetic content of the stones – in the same way modern trains float above their electromagnetic tracks and helped them lift stones heavier than 10 tons. They used a floating substance, the same monoatomic gold. The roads connected pits to houses, terraces, workstations and ceremonial centers." (pg. 55- "Anunnaki Gods No More" – Lessin)

28. First Base Way Station was Mars, then after Mars was affected by a Nibiru entry, they moved it to the Moon

29. Nephilim – Anunnaki mine bosses

30. Enki's first 2 children were not named Adam and Eve- they were named Adapa and Titi, then they mated and had **Ka-in and Abael**

31. Ka-in killed Abael because Enki found more favor with Abael's offerings of lambs for meat and wool and not Ka-in's grains and fish-filled water canals

32. The beardless Indians of the Western Hemisphere owe their decendence to Ka-in.

33. Marduk was sentenced to die in the King's Chamber in the Great Pyramid

34. After the Flood only about 1,000 humans were found hiding in mountain caves. Mud was everywhere. Mud destroyed the rocket terminal at Sippar. The only surviving edifice after the Flood was the Baalbek Landing Platform. The humans then began to grow crops. Plows were invented. The wheel was invented.

35. Ningishzidda constructed the Great Pyramid

36. Found gold in alluvium at Lake Titicaca in Peru, and also descendants (beardless) of Ka-in.

37. The pyramid was originally a communication device to Nibiru

38.   The Sphinx face was originally of its builder – Ningishzidda

39.   At the top of the pyramid were installed "pulsating crystals and the Gug Stone, a capstone of electrum, to reflect a beam for incoming spacecraft."

40.   Dead Sea is "dead" because of nuclear bombs from 2024 BC, which contaminated it with radio-activity

41.   Marduk was anti-women

42.   Enlil "branded" humans during his war with Marduk by ordering their foreskins cut off (similar to his own phallus, and now 'loyal' to Enlil)

43.   The "Face" on Mars is that of Alalu – it exists today

44.   The Maya are direct descendants of Cain

45.   Tephera are rocks that have been exposed to nuclear radiation – blackened and gravel-like

46.   In the Sumerian language E.DIN means "Home of the Righteous Ones". Obviously E.DIN and Eden are one and the same

47.   Enki means - Lord (En) of Earth - (Ki)

48.   Enki had 2 sons – Marduk (Ra), then Ningishzidda (Thoth)

49.   One of Enlil's sons was Ishkur, who some authors believe was the Yahweh or Jehovah of the Old Testament.

50. When the Anunnaki came to a huge region of marshes - Enki built dikes and irrigation works and created a "Raised Land" or Egypt.

51. The Anunnaki were a human race with human bodies and an exuberant sexuality. They lived in cities and were highly individualistic. Power was important to them.

52. They were highly intelligent and had mastered inter-planetary travel.

53. They established a space station on Mars, and possibly the Moon as well.

54. They were sophisticated genetic engineers and had mastered cloning. They could manipulate energy fields, had medical beams, guiding instruments, were able to forecast planetary movements, disequilibrium, and catastrophes.

55. They had near immortality.

56. They had high tech machines called - "ME's"

57. They had nuclear energy at least 500,000 years before us

58. Had a population of 300 "Igigi" ("those who see and orbit") on Mars

59. Noah was called "Atra-Hasis" or "Exceedingly Wise"

60. Abel (son of Adam) was taught shepherding by Marduk, and Cain was taught farming by Ninurta. Enlil showed favoritism to Cain. Sibling rivalry ensued.

61. Ninurta (son of Enlil) is named "The Warrior of Enlil" - because he was the commander of

Enlil's armies, and developed new weapons and tactics.

62. Marduk was the head of the Igigi, and the Mars Mission Center

63. The last dynasty of Akhenaten and Nefertiti understood the sophistication of stargazes, sacred geometry, free energy and antigravity.

64. According to some beliefs - the Reptilian race called - *Shemsu Hor* - formed the *Brotherhood of the Snake*- these beings are depicted in Egypt in the Temple of Hathor

65. It is believed that the Anunnaki communicated with humans through radio waves emitted through the use of crystals, which the Sumerian tablets referred to as "transference crystals"

66. The Anunnaki left Earth between 610-560 BCE.

67. Cain's descendants eventually became the Aztecs, Mayas and Incas (sounds like Enki, doesn't it?). They were aggressive and warlike, and practiced cannibalism.

68. Anunnaki life spans were 120 sars which is 120 x 3,600 or 432,000 years. According to the Kings List - 120 sars had passed from the time the Anunnaki arrived on Earth to the time of the Great Flood

69. In Australian lore, a "bullroarer" was an object making a roaring wind sound. Couldn't this be a metaphor for a UFO?

70. Eridu, the first Anunnaki city was established around 445,000 years ago when the Earth was gripped by an Ice Age

71. The Anunnaki miners mutinied around 300,000 years ago

72. Bitumens and asphalts (petroleum products) were prevalent in Mesopotamia and fueled the kilns, crucibles and furnaces

73. Enlil was feared by humans. Evidence of this is demonstrated by the lack of his appearance in the arts and legends of Sumer

74. Enki's home was Eridu. He is described in the arts and legends of Sumerians as a benevolent god and is depicted on Sumerian seals and monuments. He is aligned with the sea and rivers, and is sometimes described with a fish tail (similar to a mermaid), although the author is unsure of this description

ANUNNAKI GOD

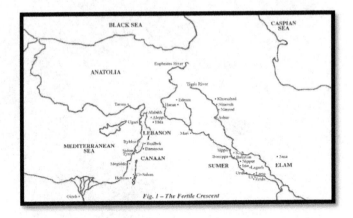

Fig. 1 – The Fertile Crescent

THE FERTILE CRESCENT - ANCIENT
SUMER AND CITY LOCATIONS

75. Ninhursag lived in Shuruppak and is described in matriarchical terms as the "heavenly mother". Her sacred symbol was an umbilical cord.

76. There were seven ruling gods in Sumer. The Menorah (Jewish sacred candelabra) has seven arms, there are seven days in a week, and the Earth was created in seven days. Coincidence?

77. Nannar, the eldest son of Enlil, lived in Ur (capital city), and his Semitic name is Sin.

78. Shamash was the Sun God connecting Heaven and Earth. He ruled over Lebanon

79. Ishkur, the youngest son of Enlil, was usually depicted holding a forked lightning bolt (as in Thor)

80. Adad is also known as Yahweh (YHWH) in Hebrew, may have been the actual father of Isaac

I find it interesting that Cain appears to be born of Eve and God, as opposed to Eve and Adam. Eve is quoted in Genesis saying - "I have added a man-child with the help of the Lord", and when Abel is born, Genesis says "next she bore his brother Abel" - this clearly suggests that Cain was offspring of the Lord God, but Abel was not. After Cain kills Abel, he is banished of Eden to "the land of Nod". Cain then marries his sister Awan when Adam reaches 200 years old. He is then reportedly accidentally killed by Lamech, thereby terminating his line of descendants, in favor of Seth. Seth is a natural born son of Adam and Eve. Seth then marries his sister and gives birth to Enosh (Enoch). Enoch is unequivocally favored by the gods and taken "on their wings" (in a spaceship) to the heavens. Here he is anointed and given "clothes of glory". He spends 60 days upon the spaceship, and then returns to Earth.

In "Flying Serpents and Dragons", R.A.Boulay describes the Anunnaki as having reptilian features (scaly skin, horns and tails), which I cannot confirm from my research but find interesting. I have heard fables depicting Jews as having "horns" on their heads, thinking that it was a reference to the evil serpent from Adam and Eve. But Mr. Boulay describes the religious ceremony of a "bris" which is, in modern times, a circumcision, as follows:

"As a sign of loyalty and a way to identify his (Yahweh's) supporters, and to remind them that they are directly descended from a reptilian god, the shedding of the foreskin was introduced at this time in the rite of circumcision. Symbolically, it represented the reptile's sloughing of his skin as the act of renewing his life." (Boulay - Flying Serpents and Dragons)

ENKI'S
CONTRIBUTIONS 1

1. Created humans
2. Discovered the gold mines in Africa
3. discovered a method to filter gold from the water of the Persian Gulf
4. Created the first city of Sumer - Eridu
5. The firstborn son of Anu

Enlil's Destruction:

- Released thermo-nuclear weapons against Marduk in 2024 BC
- Stormed Enki's gold mines in South Africa and abducted the workforce, thereby halting the gold production
- Severed the communication between Earth and Nibiru
- Added the extra chromosomal genes needed for procreation to Adam who could not procreate
- Destroyed Sodom and Gomorrah

ENKI'S
CONTRIBUTIONS 2

nki has a special relationship with Noah (also called Ziusudra). Enki warned Noah about the flood before it came, and instructed him to build an ark – no, not a huge wooden boat......a submersible, or a submarine. The ark was never a wooden boat, it is impossible to build such an enormous wooden boat, even with today's shipbuilding technologies. The ark could also never hold two of every animal species on it! This is absurd! Imagine tens of thousands of species of animals – worldwide?? Impossible. I believe the flood was real, but it was a regional, not world-wide event. I believe it happened only in the fertile crescent – modern day Iraq, and possibly other countries in the Middle East.

Enki's autobiography is written on 12 tablets. This text is called the "Eridu Genesis". In it, the "ME" is described as a data-disc that encoded all aspects of civilization on Earth. This text sheds light on Enki's attempts to bear a child with his half-sister Ninharsag, his relationships with goddesses and humans alike, and the consequences he faces as a result of this. The *Atra-Hasis* text discusses Anu's efforts to calm the relationship between Enki and Enlil by dividing up the Earth between them.

Excerpts from the Enki Text: (taken from "The Lost Book of Enki" – Sitchin)

These are the exact words dictated by Enki to a scribe known as Endubsar-. I'd like the reader to read Enki's words which were written thousands of years ago and determine for yourself if you believe this is real, or a complete myth. And if this is a myth, how could the author have known about these events? Additionally, the text was *written in the form of poetry*, verse by verse. Author's notes are in parenthesis :

"I am your Lord Enki. I am much distraught by what has befallen Mankind. My hands are not clean, not since the Great Deluge (The mythical, and true, Flood) had such a calamity befallen the Earth and the gods and Earthlings. But the Great Deluge was destined to happen, not so the great calamity (nuclear explosion). This one seven years ago..."

ENKI

ENLIL

THE FIRST TABLET

"By Enlil and Ninharsag it was permitted; I alone for a halt was beseeching. Ninurta, Enlil's warrior son, and Nergal, my very own son, poisoned weapons in the great plain then unleashed.

The Great Deluge was destined to happen; the Great Calamity of the death-dealing storm was not. By the breach of a vow, by a council decision it was caused; by Weapons of Terror was it created. By a decision, not destiny, were the poisoned weapons unleashed; by deliberation was the lot cast.

A great planet, reddish in radiance; around the Sun an elongated circuit Nibiru makes. In the cold period the inner heat of Nibiru it keeps about the planet, like a warm coat that is constantly renewed.

(Nibiru description) – The nation of the north against the nation of the south took up arms (sounds like our Civil War?). Then a truce was declared; then peacemaking was conducted. Let there be one throne on Nibiru, one king to reign over all.

In the atmosphere a breaching ozone depletion has occurred; that was their finding. Volcanoes the atmosphere's forebears, less belching were spitting up! Nibiru's air has thinner been made, the protective shield has been diminished! (if a myth, how could the author know such descriptive atmospheric details?)

What atmospheres they possessed by observation and with celestial chariots (UFO's) intensely were examined. In the councils of the learned, cures were avidly debated; ways to bandage the wound were urgently considered. One was to use a metal gold was its name. On Nibiru it was greatly rare; within the Hammered Bracelet (our Asteroid Belt) it was abundant. It was the only substance that to the finest powder could be ground; lofted high to heaven, suspended it could remain.

(Nibiru Atmosphere Issue) – One was to use a metal gold was its name. On Nibiru it was greatly rare; within the Hammered Bracelet (Asteroid Belt) it was abundant. It was the only substance that to the finest powder could be ground; lofted high to heaven, suspended it could remain....In the land strife was abundant; food and water were not abundant....Let celestial boats be constructed, he (Alalu – the first of the Anunnaki to land on Earth) decided, to seek the gold in the Hammered Bracelet, he decided.

(Problem with the Hammered Bracelet) – By the Hammered Bracelets the boats were crushed; none of them returned. (this text was written more than 5,000 years ago – if this were a fake – how could the author know about the "Hammered Bracelet" anyway?)....To snow-hued Earth Alalu set his course; by a secret from the Beginning he chose his destination.

(on Gold) – Nibiru's probing chariots like preying lions they devoured; The precious gold, needed for surviving, they refused to dislodge…Gold, much gold, the beam (tractor beam?) has indicated….

(Alalu's Earth Issues) – As a cloak for protection he the Pulser and the Emitter put on….He put on an Eagle's (interesting that the symbol of U.S. strength is the Eagle) helmet he put on a Fish's suit (scuba gear?)….alone on an alien planet he stood…It breathed the planet's air; compatibility it indicated!.....The brightness outside was blinding; the rays of the Sun were overpowering!...a mask for the eyes he donned…

(Description of Earth) – The waters with fishes were filled! For drinking the water was not fit, Alalu greatly disappointing….trees with fruits were laden…Sweet was the smell, sweeter the taste was!..Alalu greatly it delighted….Into the pond the Sampler he lowered; for drinking the water was good!...A coolness did the water have, a taste from Nibiru's water different….A hissing sound he could hear (a snake not found on Nibiru), a slithering body by the poolside was moving! (Interesting that Alalu encountered a snake on Earth, which he had never seen on Nibiru, and that snakes are the symbols of DNA and are feared on Earth)…The shortness of the day Alalu pondered, its shortness amazed him (days on Nibiru are obviously longer than on Earth)….Kingu (the Moon), the Earth's companion, he now beheld…

(On nuclear weapons) – Of the Tester its crystal innards he removed, from the Sampler its crystal heart he took out; Into the Speaker he the crystals inserted, all the findings to transmit....With Weapons of Terror (nuclear device) a path through the Bracelet he blasted! (voyage from Nibiru through the Asteroid Belt to Earth)...the Water Thruster to prepare (used water to propel the asteroids away from the chariots?)

(On Mars) – There is water on Lahmu....snow white was its cap, snow white were its sandals...Reddish hued was its middle, in its midst lakes and rivers were aglitter!...The waters were good for drinking (interesting!), the air was insufficient....

(On Eridu, the first Earth colony) – Eridu, Home Away from Home, is established in seven days (isn't it interesting that the Bible says God created the Earth in seven days! Coincidence??)...**Let this day be a day of rest**; the seventh day hereafter a day of resting always to be! (coincidence?)

(On Gold Mining in South Africa) – Where the landmass the shape of a heart was given, in the lower part thereof, Abzu (So. Africa), of Gold the Birthplace, Ea (Enki) to the region the name gave...From Earth's bowels, not from its waters, must the gold be obtained....(Enlil asks for a plan to mine correctly) – Proof of the golden veins is needed, a plan for success must be ensured!...Let him (Enki) proof obtain, a plan put forward...(Anu comes down to Earth) – his chariot splashed down....With hesitation Alalu stepped forward, with Anu he locked arms....Let the Edin (Eden) be....The Commander of the Edin let me (Enki) be, let Enlil the gold extraction perform! (so Enki is asking Anu permission to have Enlil do the mining)....Let us draw lots! Anu said. By the hand of fate let there be a decision! (Enki loses the draw of lots, and Eden belongs to Enlil) – Ea's eyes filled with tears of Eridu and the Edin he wished not to be parted.

(Machines used for mining) – An Earth Splitter with cleverness Enki designed, on Nibiru that it be fashioned he requested. Therewith in the Earth to make a gash, its innards reach by way of tunnels...Power beams the surface he flattened. Great stones from the hillside the heroes quarried and to size cut. To uphold the platform with sky ships they carried and emplaced them.

BAALBEK PLATFORM

(It is my belief that the platform Enki is referring to is the Baalbek platform which is the largest monolith in the world – see photo)

Laarsa and Lagash by Enlil were constructed, Shurupak for Ninmah he did establish (these are the 3 cities established by Enlil in Iraq?)…Those who on Earth are shall as Anunnaki be known, Those Who from Heaven to Earth Came! Those who on Lahmu are Igigi shall be named, Those Who Observe and See they shall be!

(Enki takes nukes from Alalu) – In the nearby cave Enki seven Weapons of Terror has hidden, From Alalu's celestial chariot he had them removed.

(The "ME" is a disc or device that is encoded information) – On ME's were the secret formulas of Sun and Moon, Nibiru and Earth and eight celestial gods recorded. With evil purpose Anzu the Tablets of Destinies seized. There in the Landing Place, rebellious Igigi for him were waiting, to declare Anzu king of Earth and Lahmu they were preparing! (a mutiny of the Igigi which starts a war)

Lightning darts Ninurta (son of Enlil) at Anzu directed; the arrows could not approach Anzu backward they turned. For his son Enlil a mighty weapon fashioned, a Tillu missile it was (and thereafter Ninurta defeats Anzu)

(Delivery of Gold) - In Bad-Tibira (in Iraq) they were smelted and refined, by rocket ships to Lahmu they were sent: In celestial chariots from Lahmu to Nibiru was the pure gold delivered. (but the mining was difficult, the workers complained and mutinied! Enki came up with an idea to create a "Lulu" – or a primitive human made of Cro-Magnon hominid combined with Anunnaki DNA) – Let us create a Lulu, a Primitive Worker, the hardship work to take over. The Being that we need, it already exists! All that we have to do is put on it the mark of our essence, thereby a Lulu, a Primitive Worker, shall be created! So, did Enki to them say…

(On creating man) – The name of Adamu was given by the Anunnaki as the name of the first *men*, as opposed to only one *man*. Adamu meant "One Who Like Earth's Clay Is" (a reference as described clearly in the Bible – that man was made from clay. Then, as is practiced by Jews worldwide today, the foreskin was removed after birth – "Then in the male part of Adamu an incision is made, a drop of blood to let out". Enki references our 22 DNA chromosomes like "two entwined serpents." (Ningishzidda, the essences) separated and arranged like twenty-two branches on a Tree of Life and were the essences". But the Anunnaki either knowingly restricted humans' ability to reproduce – "the ability to procreate they did not include!". Here, another Biblical reference to Adam's rib being used to create Eve – "From the rib of Enki (not Adam) the life essence he extracted, Into the rib of Adamu the life essence of Enki he inserted; from the rib of Ninmah (not Adam) the life essences he extracted, Into the rib of Ti-Amat (Eve) the life essence he inserted". The Anunnaki also restrict the humans' mortality – "with wisdom and speech they are endowed: with Nibiru's long lifetime (which is tens of thousands of years) they are not".

(On Cain and Abel) – I always wondered who Cain mated with in order to reproduce when he was abolished from the Garden of Eden, but it is explained in the Eighth Tablet – "With his sister Awan as a spouse Ka-in from Eden departed".

(On the flood) – Since I was a child, I could not wrap my arms around the idea of Noah collecting two of each animal in the world and holding them on a boat. How could this be? Lions, rhinos, giraffes, etc.....but the Tenth Tablet describes that Noah had collected the "male and female essences and life eggs they collected" – Noah had kept their DNA, not the animals themselves! "A box of cedarwood in his hands he held, the life essences and life eggs of living creatures it contains". Additionally, the Tenth Tablet describes the boat as a "submersible" or a submarine! Not a 400-ft long boat as I had been led to believe.

(On the pyramids) – the Anunnaki built the pyramids, not Jewish slaves. "By the Anunnaki, with their tools of power (lasers?) were its stones cut and erected….With galleries and chambers for pulsating crystals….Of electrum was the Apex stone (capstone/top) made, the power of all the crystals to the heavens in a beam it focused". The pyramids acted as a machine powered by the crystals – "Inside eerie lights began to flicker, an enchanting hum the stillness broke, Outside the capstone all at once was shining", "A pulsating beam that far and high reaches it emits". "The Age of the Lion let it announce (reason for building the Sphinx), as Marduk who now claims Earth as his alone, demands that the face of the Sphinx be his. However, after Marduk had warred with Ningishzidda, the face of the Sphinx was replaced with the face of Asar (Marduk's son)

WAR

"The Lord rained upon them (the sinful cities of Sodom and Gomorrah) from the skies, brimstone and fire" - The Bible.

U, THE SUPREME ANUNNAKI leader had decided to promote his son Enlil as the commander over the gold mining operations on Earth and to oversee its shipments to Nibiru. Anu demoted his son Enki as "lord of Earth", having his only responsibility with the mining operations in South Africa - in the Ab.Zu. Enki was outraged.

The Anunnaki gods fought over land. Enlil and his family went to war with Enki. It seems to have stemmed from birth rite - Enki was Anu's first born son but was jealous of Enlil who was the "legal" heir to Anu and his throne. Enki was resentful over the fact that he was given the African lands as his territory (which included Egypt), while his brother had been given the Middle East.

Enki's symbol is the snake. We see it in medicine - the snake is the Hippocratic symbol of healing. It also can be seen as a representation of DNA.

Enlil's symbol is the eagle. The power of flight and strength. So, with this in mind, I find the following quote by Alan Alford fascinating-

"There is no doubt that both South American and Mesoamerican cultures preserve the records of an Enlilite victory over Enkiite gods in the relatively recent past. The national emblem of modern-day Mexico is an eagle grasping a snake with its beak and claw, the snake being an Enkiite symbol." (pg. 542)

The Anunnaki nuked their spaceport in the Sinai Peninsula in 2024 BC. Nergel, the son of Enki, initiated this act. Nergal had disagreed with his father Enki and *sided with* Enlil. Ninurta, Enlil's son, joined Nergal in this nuclear act. Samples of blackened rocks taken from the area of this nuclear event, have a high ratio of the uranium 235 isotope. How is that possible if there (supposedly) was never a nuclear explosion in all of the Middle East?

In my prior book, ***"Adam = Alien"***, I discussed the evidence of nuclear destruction in the ancient cities of Sodom and Gomorrah. Words like "sulphurous fire" raining from the heaven, and the Lord casting down a "thunderbolt" upon the city and setting it on fire, speak to an actual nuclear event taking place. The thunderbolts were cast from the "shekinah" or the Hebrew word for an aerial chariot, which "descended to work the destruction of the cities" (Hebrew *Haggadah*). Lot's wife who was warned not to look back, "beheld the Shekinah" and was vaporized by the heat. It is my opinion that the Dead Sea formed as a result of this explosion which destroyed all living things in its path. Yahweh, the Hebrew god, may have been the culprit of this event. He is shown to be a jealous and vindictive god throughout the Old Testament. Yahweh is also known as Adad or Hashem.

How did the people of Israel communicate with God? R.A Boulay believes that when they "set up an altar", they actually were creating a localized transmitting station. Cylinder seals from 3000 BC depict reed huts with strange "antennae-like projections on the roofs with round eye-like objects attached. These huts were portable, by land or even by boat!" The Ark of the Covenant was probably the actual transmitter of information and proved to be so powerful that it killed anyone who ventured too close. Two sons of Aaron were actually killed by a sudden discharge of electricity produced by the Ark. Only the Levi tribe were allowed to prepare the Ark, apparently with protective clothing. Even today, the Levites are the only Jews permitted to open and close the Ark in today's contemporary synagogues.

It can be suggested that the followers of Enlil are the sons of light, while the followers of Enki are the sons of Darkness. Enlil ultimately wins the war, and Enki must pay a "penalty". Enki must be what the Bible suggests as "The Devil" or "Satan". In my opinion, the voice of

God in the Bible is the voice of Enlil. As Enlil is a jealous God, and expects blind obedience to him and his rules, our ultimate prize is going to "heaven". If "heaven" is Nibiru, and when we die it is really the end of us, then demonstrating blind obedience to Enlil is a deception of monumental proportions.

Marduk was known as "Ra" and Ningishzidda was known as "Thoth". Marduk, Enki's firstborn son, claimed that he, and not Enlil's firstborn son Ninurta, should inherit the Earth. This bitter conflict between Marduk and Ninurta included a series of wars here on Earth which led to the use of nuclear weapons (see my chapter on Vitrification in *"Adam = Alien"*.) This nuclear annihilation resulted in the demise of the Sumerian civilization. Interestingly though, it was his father Enki who, alone, stood in opposition to the use of such "forbidden weapons".

One of these forbidden weapons may have been the "Tablets of Destinies" which is discussed in detail in the Sumerian tablets. At Nippur, the "mission control" center of the Anunnaki, were kept celestial charts, orbital data panels and, in a dark glowing chamber, the Tablets of Destinies. The younger brother Enlil had been in control of Nippur and the Tablets. However, another Anunnaki god named Kumarbi stole the Tablets and escaped. Anu, having discovered that Kumarbi stole them, declared war. Enlil's son Ninurta, armed with a spacecraft with nuclear and laser-guided weapons, attacked Kumarbi, seized him and decapitated him.

Enki founded the "Brotherhood of the Serpent". It became a secret society in ancient times and has continued to this day. It was originally dedicated to the "dissemination of spiritual knowledge and the attainment of spirituality for all spiritual beings including humans." (Pruett page 166). In Hebrew, the word for serpent is *Nahash*, which means "he who knows all secrets". But Enlil guided the Brotherhood away from spiritual growth to spiritual decay. It has become an agent for "the suppression of truth and knowledge for mankind" (Pruett pg. 166)

The Brotherhood has forcefully contrived a New World Order with a centralized banking system that has created two distinct classes – the super-wealthy who control 99% of the world's economy, and the poor (almost 7 billion people) who combined control only 1%. The Brotherhood has produced an economy which produces artificial cycles of inflation and deflation. The Fed controls these cycles and exists to manipulate the control of money into the hands of The Brotherhood.

Honest presidents like Lincoln and John F. Kennedy were assassinated when they tried to abolish the Fed. And there is ample evidence to prove this.

ACTUAL DRAWING OF NIBIRU ON ROCK.
NOTICE THE CROSS TO DELINEATE IT AS THE
"PLANET OF THE CROSSING"

ANUNNAKI CYLINDER SEAL SHOWING GODS
WITH 10 PLANETS IN BACKGROUND. 10
PLANETS, NOT 9. THE TENTH BEING NIBIRU.

PROOF OF ANUNNAKI

In March 2013, there has been a startling discovery near the Iraqi city of Ur. It was a **secret joint collaboration** between British and Iraqi archaeological teams. Tel Khyber is in the southern province of Thi Qar (approximately 200 miles south of Baghdad). The discovery was not announced publicly. It makes me think that the US invasion of Iraq in 2003 had, among its objectives, to do with the discovery of a portal, and evidence of Anunnaki civilization and additional written documentation of our real history. Interestingly, the name "Ur" means "to shine" in Hebrew. In the 1930's Ur was excavated by Sir Charles Leonard Woolley, son of George Herbert Woolley. Sir Charles analyzed the effects of the biblical flood at Ur, and proposed that it was a *local* event, and not a worldwide event. The author agrees.

The discovery was amazing!

It was a huge complex of great importance. The walls of the complex were *nine feet thick*. There were separate buildings organized in a square.

There are also interesting connections between Cain's banishment from the Garden of Eden, to his development in Latin America, especially Mexico and Central America.

Sitchin remarks that Cain's Egyptian name was "Ka'in" which backward is "In'ka" or "Inca". Enoch as well, may have journeyed to Mexico as evidenced in the name "T*enoch*titlan". Titlan meaning "city of" Enoch. This is the capital city of the Incas and Mexico.

ACTUAL HIEROGLYPH OF A STARGATE - A HUMANOID EMERGING THROUGH A PORTAL. AN INCANDECENT TUBE IS SEEN THROUGH WHICH THE FUGURE OF A MAN IS EMERGING

CHAPTER 5

THE ADAM CODE

THE ADAM CODE

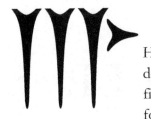HE SUMERIAN TABLETS describe how Enki created the first human beings. It states the following -

"Male and female created he them... and called their name Adam."

It is of critical importance to understand that Adam was not a personal name at all. Adam had to do with the Earth and Enki's DNA mixture of Anunnaki and primates. The Sumerian word for Earth was- *adamah* - which was the origin of the name Adam. The first-century Jewish writer Flavius Josephus describes Adam to be the color of red, because he was *"compounded out of the red Earth"* - This was a reference to the clay jars which held the DNA mixture of humans as depicted in various Sumerian cuneiform tablets. Even more interesting is that the Hebrew word for red is - *adom*. The Vedic word for mighty is - *hu*, and the term *hu-mannan* identifies as "mighty man".

In the Bible Adam and Eve are originally described as being naked. This nakedness has nothing to do with anything related to sex or their supposed guilt about being innocent and their sexuality. In all of the Sumerian Tablets reliefs, the humans are depicted as being naked - this is more of a description of their servitude and subordinate status as compared to their gods, the

Anunnaki. In fact, Adam and Eve's nakedness had nothing whatsoever to do with sex, as there is no mention of any physical contact between them. As Gardner posits -

> "It is commonly believed that the Christian term 'Original Sin' had something to do with Adam's and Eve's sexual behavior, but this is a church promoted absurdity. To the point where Adam is banished from the garden, there was no mention whatever of any physical contact between him and Eve. The eventually determined sin was that Eve (a mere woman in the church's eyes) had seen fit to make her own decision: a decision to disobey Enlil in favor of Enki's advice, a decision to which Adam conceded and a decision which proved to be the correct one. In practical terms, Eve had committed no sin at all because the interdict concerning the Tree of Knowledge had been placed on Adam alone, which is why only he was exiled."

(Gardner – pg. 130)

The concept of Original Sin had been developed and promoted by St. Augustine, in an attempt to support the church's sexual paranoia. The concept of the serpent "tempting Eve" lead to the myth of Satan. Enki has long been connected to the symbol of the serpent and Satan. But nowhere in the Hebrew Bible is Satan ever discussed. It is a wholly invented myth originating by the Church and its bishops in the post-Jesus era. This is one of the true *myths* of the church's supposed *history*. The God-Satan conflict was representative of the Enlil-Enki conflict which was representative of the battle between Light and Darkness. The Cain-Abel conflict is also symbolic of the

Enlil-Enki conflict, in that they both had different fathers. Cain was the son of Anunnaki blood and Eve, while Abel was the son of Adam, a homo sapien, and Eve. Cain then emerges as the victor after killing Abel, since he had been the more advanced and stronger of the Royal Seed.

ADAM AND EVE. THE SNAKE REPRESENTS
ENKI SPEAKING TO EVE.

CREATION OF MAN

Here Enki describes the process – "In a crystal vessel Ninmah an admixture was preparing, the oval of a female two-legged she gently placed (insertion of the zygote into a female uterus). The result – "his foreparts (foreskin) like of the Earth creatures were"…the baby had the foreskin of primates, "his speech only grunting sounds was!" (incapable of speech). So, they kept trying…on and on….they changed the method of using crystal to using clay….

Then they inserted the zygote into a female Anunnaki woman, "Perchance the right admixture in the wrong womb was inserted"…then…"He slapped the newborn on his hind parts; (rear end) the newborn uttered proper sounds!". "Dark black his head hair was, Smooth was his skin, smooth as the Anunnaki skin it was". The penis still had a foreskin, but the child could speak. "Adamu (Adam) I shall call him! Ninmah was saying. One who Like Earth's Clay Is, that will be his name."

"*In our image and likeness*". In my previous book - **"Adam = Alien",** this quote was on the front cover. This direct quote from both Old and New Testaments drives home my thesis. The word "*our*" is plural and not singular. If God were the one, singular maker of us, why would he be referred to in the plural? It should say "In **my** image and likeness"....wouldn't it? Also, if God was an ethereal, not corporeal being, how could we humans be corporeal? How could we have a body? The scribes of the Bible knew that God was plural....that's why they kept in this verse. They knew that the truth, that our makers were *"gods"* and not God. Our makers had physical bodies, and when they made us, our image or form was, in fact, similar to theirs....hence the words *"our image and likeness"*.

It is interesting to note that humans have two blood types - *Rh* positive and *Rh* negative. The Rh-positive types are associated with hominids, while the Rh negative are associated with a direct line of Anunnaki. Apes contain 24 chromosome pairs, while humans contain 23. What happened to one chromosome pair? The removal of one pair would have created too much damage to the transmutation process, so the only logical solution left was to *splice* two chromosomes together.

Apes don't get diabetes, but humans do. Why? It is because this gene-splicing process caused problems that only affect humans. Apes would have eradicated this disease over time, but humans get diabetes as a direct result of the Anunnaki's imperfect gene-splicing process. Humans carry over 4,000 genetic disorders. Apes carry almost none. Where these disorders put into our code, or carried on as a result of the gene-splicing mistakes? Lloyd Pye points out the difference between Darwin's theory of evolution, and what the fossil record actually reveals to be the truth -

"Sea worms did not and do not become fishes, fishes did not and do not become amphibians, amphibians did not and do not become mammals. In every case, the differences between critical body parts and functions (internal organs, digestive tracts, reproductive systems, etc.) are so vast that transition from one to another would require dramatic changes that would be easily discernible in the fossil record. What the fossil record actually reveals is that every class, order, family, genus or species simply appears, fully formed and ready to eat survive and reproduce." (Pye - Everything You Know is Wrong).

DNA's shape makes it a perfect receiver-transmitter due to its crystalline structure. **It can store more than 100 trillion times more information than any device known to man.** We have 120 billion miles of DNA in our bodies. However, more than 95% of it appears to have no known function. Perhaps when the Anunnaki designed our DNA, *they purposefully turned off* 95% of its function? Was this because they wanted us only to perform only the tedious labor of digging their gold and building their pyramids?

The problem with Nibiru's atmosphere is described by Hardy - "Nibiru's atmosphere - thought to be protected and maintained by particles spurted out by volcanoes- was dangerously dwindling. (Let's note here that this is scientifically sound, because the ashes and particles of an eruption - by rising to the upper atmosphere- end up screening UV rays and thus lowering the temperature on the Earth's surface). In this perspective, Alalu's (one of the main scientist's) solution to the problem had been to use nuclear blasts to force the eruption of volcanoes, but the effect on Nibiru's atmosphere had been short-termed. There was another scientific possibility, but it was out of reach for them: Nibirian scientists knew that they could spray gold particles in the upper atmosphere to create a protective veil encircling it and retaining the atmosphere captive; however, there was not enough gold on Nibiru." (Hardy - pg. 26)

The main Hebrew God Hashem may actually have been Enlil's youngest son - Ishkur. How do we know this? According to Alan Alford, Ishkur was anti-Babylon and anti-Egypt. He had a violent streak and was very jealous. He has frequently been depicted carrying a trident, thunderbolt or forked weapon. He quickly lost his temper.

It is startling that we are confronted with the realization- one that our Judeo-Christian heritage has been hiding for generations- that the Anunnaki are in fact, *our gods*. They are the ones who created us using their own genes, and they themselves were a human race!
But why would they, being so scientifically advanced, let their atmosphere dwindle to the point of planetary extermination? It doesn't make sense. We can, however, see ourselves doing the same thing! Our fossil fuels and pollution are doing the same thing to our atmosphere - destroying it.

The word Anunnaki means "Those who from Heaven to Earth came". They are described pictorially as "DIN.GIR" Sitchin describes the word "GIR" as a "rocket". Take a look at the symbols used to represent the Anunnaki.

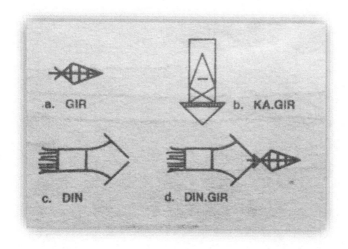

Not only does this show a diagram of a rocket, but even goes further to delineate a rocket inside of a chamber. The spaceship has a landing craft docked into it, similar to our lunar module being docked with Apollo 11. Isn't it interesting that the Sumerian word "shem" translates to a "sky vehicle", and the Hebrew word "Hashem" refers to "God in Heaven"....

EN.LIL which translates to "Lord of the Wind or "Lord of the Command" controlled the city of Nippur. Nippur's ruins still exist today (see Photo), 100 miles south of Baghdad. EN.KI translates to "Lord of Earth". His city was Eridu, which is located at the mouth of the Tigris and Euphrates rivers in the Persian Gulf. Enki created man. NIN.HAR.SAG was Enki's half-sister, who assisted Enki in the creation of man. Their first creation was called a LU.LU, which interestingly translates to "one who has been mixed".

NIPPUR RUINS TODAY

According to author Gerald R. Clark who wrote *"The Anunnaki of Nibiru"* –

"One of the most interesting finds was the Temple of Hathor atop the Biblical Mount Sinai. Within the temple was found a strange white talcum powder that was apparently the result of smelting gold as it turned out. This find let to the re-discovery of mono-atomic gold

by David Hudson. These room temperature superconductors have anti-gravity properties and have been postulated to have been leveraged to move the large stone blocks used in temple construction. Additionally, there is clear evidence that the Anunnaki of a chosen bloodline, were ingesting the mono-atomic gold in the form of conical bread cakes as depicted on Hathor's temple walls. The shorter orbital cycles on Earth were having a negative effect on the Anunnaki DNA; specifically, the telomeres were being damaged by close proximal radiation from the sun. the ingestion of mono-atomic gold has the effects of "lighting up" the human energy body as well as provisioning a bridge to other dimensions due to the missing atomic mass of the multi-pass annealing process applied to the smelting (of) gold" (pg. 39)

In discussing The Portal and its connection to the Jewish people, Clark posits-

"Thus, the Jewish people, descendants of Jacob, were in a geographic position to be promoted as the guardians of the 'portal' that was built beneath the Temple Mount in Jerusalem. This Anunnaki - only portal was used for high-level VIP travelers that visited Earth from Nibiru like Anu. This was the post-diluvial bond Heaven-Earth. Anu visited Earth using this portal and various missions to Nibiru were launched from Jerusalem's Mission Control Center....It is the author's humble opinion that the portal of Mount Horeb, similar to the one found in Uruk, is the reason behind the world's obsession with Jerusalem. Thus, protecting access to the portal, assuming it is still functional, would be of tantamount importance on Earth, as evidenced by the alliances and maneuverings between the United

States and Israel relative to their perceived enemies." (this author agrees!)

DNA is a complete mystery....we know its physical makeup, essentially what it looks like, and it's basic (and I use that term lightly) function. It's the hidden gem of life. The elixir, the grandiose design....but what you may not know is that DNA not only looks like a wave (or two intertwined snakes), it performs like a wave. Like any wave, it creates force – a gravitational force. In "The Synchronicity Key", David Wilcock, a multi-disciplinary author extraordinaire, discusses one particular scientist named Dr. Sergey Leikin, who has studied DNA in a new way-

"In 2008, Leikin put various <u>types</u> of DNA in ordinary salt water and tagged each type with a different fluorescent color. The color-coded DNA molecules were scattered like confetti throughout the water. Much to Leikin's surprise, matching DNA molecules travelled the equivalent of thousands of miles, within their own tiny universe, to find each other. Before long he saw that entire clusters of DNA molecules have gathered together....(after ruling out electromagnetic attraction principles), gravity becomes the most likely answer within the existing energy fields known to modern science."

DNA appears to be generating a microgravitational effect that attracts and captures light. Its primary function appears to be to both absorb and transmit light. We can almost compare a DNA molecule to a miniature fiber-optic cable. The storage of light within the DNA particle is what separates healthy tissue in our bodies to stressed or diseased tissue.

History appears to be cyclical and not linear. The term – "history repeats itself" is true and can be proven. Since the dawn of "time", cycles of events have occurred for certain periods, and then repeat themselves. Books have been written proving that historical events that have happened in the past, have re-happened again. Author Michel Helmer wrote articles for the French journal called – "Les Cahiers Astrologiques" presented his theory on the cyclic repetition of events. The theory argued that the cycle was based on the Ideal Pre-Eminent Number of 25,920 and its factors. Applying this theory allowed Helmer to make exact predictions – both economic and political.

While chimpanzees have not displayed any significant mutations in their genes for tens of millions of years, humans have. If Darwin's theory of evolution was true, then humans would have descended from primates, and their genes would have mutated to become *human*. But the reverse is true - the key principle of evolution provides that successful mutations are very rare, and that natural selection processes favor simplicity over complexity. So how did we become *human*?

DNA was found on the Murchison meteorite that smashed into the ground in Australia in 1969. In a paper published in *Earth and Planetary Science* in 2008, Zita Martins wrote the following -

"We present compound-specific carbon isotope data indicating that measured purine and pyrimidine compounds are indigenous components of the Murchison meteorite. Carbon isotope rations for uracil and xanthine respectively, indicate a non-terrestrial origin for these compounds..."

Martins believed that these raw materials that are thought to have been required to create the first molecules of DNA and RNA appear to be of extraterrestrial origin (Hart - *Alien Civilizations*)

ANUNNAKI GODDESS NURTURING A CHILD
(HUMAN?) NOTE THE SPIRAL-SHAPED RODS
DEPICTING DNA

ORIGIN OF EARTH

A CCORDING TO JIM MARRS' summary of Sitchin's origin of Earth-

"Sitchin's revisionist interpretation of the Sumerian text asserts that more than four billion years ago, the planets Mercury, Venus and Mars were closest to the sun. A large watery world called Tiamat was in orbit between Mars and Jupiter. Nibiru, a large rogue planet that theoretically travels in elliptical orbit, entering our system about every 3,600 years, arrived and narrowly missed Tiamat. Tiamat cracked under the gravitational stresses. In a subsequent pass by Nibiru- in Sitchin's early works, he refers to this orb by its Babylonian name Marduk- Tiamat was cleaved in half when one of Nibiru's moons rammed into the planet.

"The collision of Nibiru's moon and Tiamat knocked a large portion of Tiamat past Mars, ripping away its atmosphere and pieces of matter of various sizes. These fragments of Tiamat remained in its original orbit, becoming the familiar asteroid belt, or the Hammered Bracelet or Firmament, as it was called by the Ancients. The great portion of Tiamat was knocked into a new orbit closer to the sun. This larger chunk, retaining much of the planet's water and carrying material from Mars, coalesced, cooled, and began orbiting between Mars and Venus, becoming Earth. It was accompanied by one of Nibiru's moons (Kingu), which was captured by Earth's gravity and became our own satellite (author

– Moon). Some say that the huge gouge out of the Earth now encompassing the Pacific Ocean is where that portion of Tiamat broke apart." (Marrs- pgs. 11-12)

According to the great Francis Crick-

"Life did not evolve first on Earth; a highly advanced civilization became threatened, so they devised a way to pass on their existence. They genetically-modified their DNA and sent it out from their planet on bacteria or meteorites with the hope that it would collide with another planet. It did, and that's why we're here. The DNA molecule is the most efficient information storage system in the entire universe. The immensity of complex coded and precisely sequenced information is absolutely staggering. The DNA evidence speaks of intelligent, information-bearing design." (Marrs – pgs. 15-16)

Lou Allamandola of NASA published his results in the British journal *New Scientist* in 1998, demonstrating that he could create complex molecules in a laboratory. All that is needed are clouds of gas in interstellar space. However, when he tried to do the same under terrestrial circumstances, it was impossible. This contradicts the mainstream concept of creating life out of sludgy ingredients in a pool. He discovered that lipids which make up the walls of individual cells where necessary to form the cell itself and could not be produced under terrestrial circumstances. The implications of this suggest that life is rampant and can be reproduced easily in these clouds of gas in space. After life has been originated it may be assumed that it was then transported by comets to other bodies. *"I begin to really believe that life is a cosmic imperative"* – Lou Allamandola – *"The Stargate Conspiracy"* – Picknett and Prince

Nibiru is called the planet of the Crossing because its orbit crosses the solar system between Mars and Jupiter.

Evidence of Nephilim on Earth is found in Arizona's Canyon de Chelly National Park. After a big washout of torrential rains, a number of skeletons emerged. All work containing this find was overseen by personnel from the Smithsonian Institution and the FBI. However, one of the Park Service employees sent an email regarding the remains of one of the graves as follows-

"male, approximately seven-foot in height, (with) six fingers and six toes"

Interestingly, the depictions of the Anunnaki and Nephilim in Egypt depict tall, muscular, thick-boned giants with beards that had six fingers and six toes.

In February 2012, a satellite from China called the "Chang'e-2" released photos of an alien base on the moon, according to www.messagetoeagle.com. It seems that China is trying to undermine NASA by confirming the fact that NASA has and continues to purposefully airbrush out sensitive details of moon photos-

"I was sent some pictures by a source who claims China will be releasing hi-res images taken by the Chang'e-2 moon orbiter which clearly show building and structures on the moon surface. He also

claims NASA has deliberately bombed important areas of the moon in effort to destroy ancient artifacts and facilities…"(www.messagetoeagle.com) – Astrada – The *"Nonsense Papers"*

Quotes-

"We find ourselves faced with powers that are far stronger than we hitherto assumed, and whose base is in present time unknown to us. I cannot say more at present, we are now engaged into entering into closer contact with those powers" – Wernher Von Braun – ex-Nazi nuclear physicist

"We cannot take credit for our record advancement in certain scientific fields alone. We have been helped by the peoples of other worlds. Flying saucers are real and spaceships are from another solar system." – Hermann Oberth – ex-NASA and rocket propulsion expert

There are "non-terrestrial ships" orbiting planet Earth, with "non-terrestrial officers" getting transferred to these ships along with "off world cargo" – Gary McKinnon – famous 2001 hacker into NASA and Defense Department computers.

Above from the *"Nonsense Papers"* - Astrada

I have also read about flights from remote US islands in the Pacific to these "non-terrestrial ships" and believe that it is quite feasible to do it.

In 1924, noted archaeologist William F. Albright discovered the ruins of an ancient city on the Jordanian side of the Dead Sea. The site was known as Babe dh-Dhra "met with some sort of disaster around 2350 BC. Archaeologists can't say for certain, but they are now considering the possibility that it may be ruins of Sodom. Human bones were found under the debris of a huge tower that had fallen, and there was evidence that the city's walls had collapsed. A layer of ash was found, suggesting that the city may possibly have suffered a major fire." Was Gomorrah close by?

Well, *"in 1973, Dr. Thomas Schaub of Indiana University in Pennsylvania and Dr. Walter Rast of Valparaiso University in Indiana discovered a site called Numeira, seven miles to the south, with pottery identical to that in Babe dh=Dhra. The city appeared to have suffered the same fate as Babe dh-Dhra, as its walls and buildings had collapsed. The site also revealed evidence of massive burning, and carbon dating set the time of its demise as the exact time as Bade dh-Dhra."* – Roberts – "From Adam to Omega"

The key question regarding evolution is this- Was this Gomorrah?

EVOLUTION

Why has man evolved so swiftly in the past 50,000 years, w hen his normal process had been virtually stagnant for the millions of years prior? If Darwin was correct, why hadn't there been hundreds of changes, and inventions during those millions of years? It doesn't add up. There must have been some injection of intense evolution into man's DNA. The "learning curve" could not have increased so dramatically by itself without any boost.

Lamarck's theory that creatures evolved because they *"wanted to"*, which is very different than saying, because they *"had to"*. Darwin agreed with Lamarck but said that *"wanting to"* was not the key component for change. Sir Julian Huxley- who was a Darwinian – stated that man has become the "managing director of evolution" – meaning that he is in complete control of his evolution. He enjoys change. He builds better homes and tools. He is not satisfied with yesterday's technology. He wants more. It is true that the top layer of the human brain, the cerebral cortex, has been the source of man's development over the past 500,000 years. This region of the brain has developed the most. The left side of the brain is the "scientist", specializing in speech, logic and reason. The right side is the "artist", specializing in shape recognition, music and telepathy! Yes, telepathy. Which is a characteristic that I believe is part of the 90% of our brains that we do not currently use enough, and that many people here on earth, including myself, have the innate ability to use on a very, very small scale. I am able to "read someone's mind", , many times without trying. I have been told that I have ESP – and I believe it. It's something I was born with and feel is part of my "makeup". But I don't really have complete control over it. It comes in waves and is inconsistent. I am also an accomplished professional musician and feel that these two characteristics are interrelated and interwoven in my "sensory makeup".

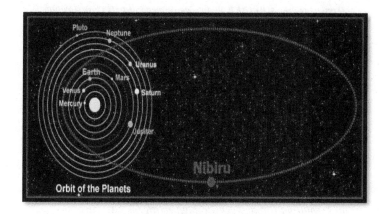

ORBIT OF NIBIRU

THE ORIGINS OF ADAM

"...How could Homo sapiens, Modern Man, appear in southeast Africa some 300,000 years ago overnight (in anthropological terms) when the evolutionary advances from apes to hominids, and in hominid species from Australopithecus to Homo habilis to Homo erectus, etc., took millions upon millions of years?" - Sitchin

"By mixing genes extracted from the blood of a god with the essence of an existing earthly being, 'The Adam' was genetically engineered. There was no Missing Link in our jump from Homo erectus to Homo sapiens, because of the Anunnaki jumped the gun on Evolution through genetic engineering." - Sitchin

Isn't it interesting that when Enki and Ninmah fashioned the first humans, the Bible referred to it as *"blowing the Breath of Life"* into Adam's nostrils? (Genesis 2:7). Ninmah was referred to as *Mami* (the Mother) - obviously the origin of our *"mommy"*. Enki was regarded as "the Serpent" in biblical verses, possibly as a phallic male symbolism, or as the opponent of Enlil who was considered the true "god" at that time. But the entwined serpent symbol has carried on down today to represent the DNA coils and the symbol of "medicine".

ALAN ALFORD QUOTES

Alan F. Alford was an independent researcher and author, who was increasingly being recognized as the world's leading authority on ancient mythology and the esoteric meaning of ancient and modern religions.

❖ How did *Homo erectus* suddenly transform into *Homo sapien* 200,000 years ago with an immediate 50% increase in brain size and suddenly had the ability to speak? Especially when after 1.2M years, there had been no progress at all....?

❖ Machu Picchu was used for the same purpose as Stonehenge- "both sites being connected to the processional change from the era of Taurus to Aries over 4,000 years ago". (Alford - pg. XVI)

❖ We are all at the mercy of the translators of the Biblical stories/myths.

❖ There are several instances where the Lord makes *physical* rather than *spiritual* appearances, suggesting that he was, in fact, made of flesh-and-blood. an example is Sodom and Gomorrah where the Lord had to descend down (physically) to assess the situation, then used a physical means (Sulphur and smoke) to destroy the people

❖ Proof of "multiple deities" in the Bible is evidenced by the fact that the Israelites pledged to worship only *one* god - Enlil- and not the *other* gods, such as Yahweh. the word "Elohim" is the *plural* of "El", or The Lord.

❖ The "clay" from which man was created can be taken quite seriously. Man's first test tube was made of clay

❖ The word "rib" from which God took to make Eve, was actually the Sumerian word "TI" which also means "life". Therefore, Eve was not created physically from Adam's rib, but rather from Adam's life essence. this basically refers to Adam's DNA.

❖ The original ancient text was called the "Atra-Hasis" named after its hero. The text discusses the whole process of the birth of Eve and sounds a lot like our current process of cloning. Eve's original name was "LU.LU" which means "the mixed one" in Sumerian - proving that she was a hybrid of the gods (humanoid) and a primitive hominid (Cro-Magnon Man)

❖ The Atra-Hasis describes how the other "gods" rebelled against their leader - Enlil (the only Hebrew god)

❖ UFO's are referred to by the Hebrews as "chariots" simply because they were land-locked and had never seen any other mobile device other than a chariot. The Egyptians whom had been sea-going and had built ships that sailed in the seas, referred to them as "boats of heaven", and the Chinese referred to them as "dragons."

CHAPTER 6

DNA

THE SPIRAL CODE

DNA-THE SPIRAL CODE

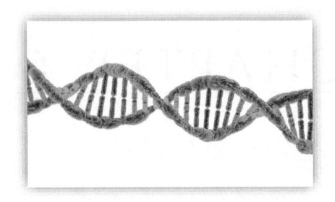

ENKI DESCRIBES DNA AS THE "Fashioning Essence" – "akin to ours it is, like two serpents it is entwined". How can this be? How can a 7000-year-old clay tablet describe the spiral nature of two DNA strands? How could the author have known this? How could this have been a fraud? It is proof positive that the Anunnaki, specifically Enki, knew the make-up of DNA. That is why there are so many depictions of snakes in the Bible. They are representatives of DNA. When his brother Enlil reminded him that they were but slaves, Enki responds – "Not slaves, but helpers is my plan!". "Not a new creature, but one existing more in our image (reference to "made in our image" again) made!" And "only a drop of our essence is needed".

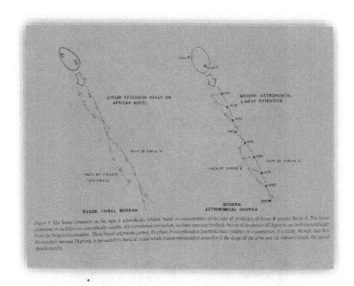

LEFT - DOGON TRIBE IN AFRICA DRAWING OF
WHERE THEIR MAKERS (GODS) CAME FROM.

RIGHT - ACTUAL ORBITS OF SIRIUS A AND
SIRIUS B. UNBELIEVABLE RESEMBLANCE.
HOW CAN THIS BE A COINCIDENCE?

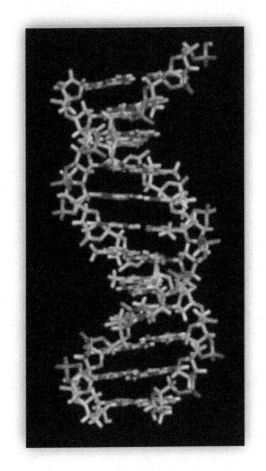

DNA

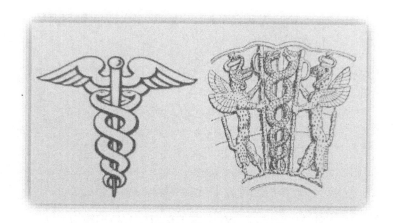

COMPARISON OF ANUNNAKI DNA SYMBOL (ON
THE RIGHT) WITH TODAY'S MEDICAL CADUCEUS
SYMBOL FOR MEDICINE (ON THE LEFT)

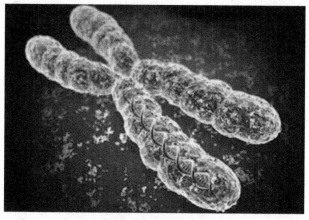

CHROMOSOME

SUMERIAN CYLINDER SEAL DEPICTING
ANUNNAKI GODS "ENLIL AND ENKI"

SPIRAL SHAPE OF THE MILKY WAY GALAXY

D N A

DNA is a complete mystery....we know its physical makeup, essentially what it looks like, and it's basic (and I use that term lightly) function. It's the hidden gem of life. The elixir, the grandiose design....but what you may not know is that DNA not only looks like a wave (or two intertwined snakes), it performs like a wave. Like any wave, it creates force – a gravitational force. In "The Synchronicity Key", David Wilcock, a multi-disciplinary author extraordinaire, discusses one particular scientist named Dr. Sergey Leikin, who has studied DNA in a new way-

"In 2008, Leikin put various types of DNA in ordinary salt water and tagged each type with a different fluorescent color. The color-coded DNA molecules were scattered like confetti throughout the water. Much to Leikin's surprise, matching DNA molecules travelled the equivalent of thousands of miles, within their own tiny universe, to find each other. Before long he saw that entire clusters of DNA molecules have gathered together....(after ruling out electromagnetic attraction principles), gravity becomes the most likely answer within the existing energy fields known to modern science."

The ratio of carbon-13 to carbon-12 (which is what the Earth is comprised of) on the Murchison meteorite is double that found on Earth, which suggests that the abundance of water is up to 20% (versus less than 1%) more likely than the compounds responsible for water production on Earth. These comparisons suggest that wherever these rocks originally came from had a greater abundance of both water and carbon compounds than our planet!

The Kepler telescope which was launched in 2009 has discovered 1,235 alien planets capable of life, 68 of which are the size of Earth, and 19 of which are larger than Jupiter.

Molecular oxygen was discovered by the Herschel Space Observatory of the European Space Agency in 2009 on Orion.

FRANCIS CRICK DEMONSTRATING THE
DNA SPIRAL

INTELLIGENT DESIGN

"If science is about following the evidence wherever it leads, then why should scientists rule out a priori the possibility of discovering evidence or supernatural design?" - Phillip Johnson, Berkeley Prof. of Law

Darwin's Dilemma:

World Perception of Life

➢ 10,000 BC – 500 BC - The Gods
➢ 500 BC - 1859AD - One God
➢ 1859AD - 1957AD - Darwin's Theory
➢ 1957AD - present - Crick's Theory

The theory of natural selection has a few major problems. First, Homo sapiens have only 46 chromosomes, whereas chimpanzees have 48. How are the two chromosomes *fused* together? It can't be explained by genetic mutation. So how did it occur? With the exception of viruses, evolution is a very slow process, taking place over hundreds of thousands or even millions of years. But the evolution from primate to Homo sapien occurred as if overnight._ How in 2,000,000 years did evolution go from tree-climbing primates_to a homo sapien sending rockets up into space while his supposed *slower* cousins still jump from tree to tree?_

Even Crick, the co-discoverer of the DNA molecule, posited that the DNA molecule was way too complex to have arisen solely through natural evolution. The "missing link" will never be "found" because homo sapiens are not the product of natural evolution, we are the product of genetic engineering as seen by the size of the human skull and the unique differences between primate and human DNA. The Sumerians, therefore, were not the beginning of a civilized human culture, they were the beginning of a series of extraterrestrial "tweaks" of homo sapiens, and the benchmark for human greater intelligence spawned on by our extraterrestrial makers.

Author Laurence Gardner voices this best in his book "Genesis of the Grail Kings" when he writes-

"In 1871, when publishing his 'Descent of Man', Charles Darwin coined the expression 'missing link' in relation to a perceived anomaly in the human evolutionary progression. There was an undeniable inconsistency in the supposed lineage which, at first, seemed like a gap in the sequence, but it was soon realized that there was no gap, simply and unexplained link".

"It took man over a million years to progress from using stones as he found them to the realization that they could be chipped and flaked to better purpose. It then took another 500,000 years before Neanderthal Man mastered the concept of stone tools, and

a further 50,000 years before crops were cultivated and metallurgy was discovered. Such was the long and arduous natural process which brought humankind to about 5000BC. Hence, by all scales of evolutionary reckoning, we should still be far removed from any basic understanding of mathematics, engineering or science - but here we are, only 7000 years later, landing probes on Mars."

Darwin's evidence was predicated on the evidence of the following finds:

1. <u>LUCY</u> - found in 1974 in Ethiopia. She lived between 3.6 - 3.2 million years ago.

2. <u>AUSTRALOPITHECUS RAMIDUS</u> - found in 1994 in Ethiopia. Lived 4.4 million years ago.

3. <u>AUSTRALOPITHECUS ANAMENSIS</u> - found in 1995 in Kenya. Lived 4.1 - 3.9 million years ago.

Additional evidence suggests the key feature in all three finds is that their skulls more closely resemble chimpanzees than men. So, in all three cases, one cannot confidently show a direct lineage to man. The missing link remains a mystery. In 1995, The New York Times, Sunday edition remarked-

"Their relationships to one another remains clouded in mystery and nobody has conclusively identified any of them as the early hominid that gave rise to Homo sapiens."

Alan Alford poses an excellent question-

"Why has Homo sapiens developed intelligence and self-awareness whilst his ape cousins have spent the last 6 million years in evolutionary stagnation?" (Alford - pg. 47)

Homo erectus survived 1.2 million years without any apparent change and then mysteriously died out. But in the process, only ONE type of Homo erectus managed to survive. That one was the Homo sapien whose brain had enlarged from 950cc to 1450cc overnight! This sudden change defies all laws of evolution. There is additional evidence that Homo sapiens co-existed with Neanderthals between 100,000 - 90,000 years ago. Addition evidence suggests that even if Homo sapiens had developed their brains to such a large extent - 50% - as a result of the need to "outsmart" or use for complex purposes....who were they trying to outsmart? Who were their competitors? What rival caused intellectual ability to be such an essential survival development tool? The Neanderthals died out, and they did not breed with Cro-Magnon Homo sapiens. The Cro-Magnon homo sapiens did not descend from the Neanderthals - they were an entirely different breed with entirely different DNA structures. It appears that modern homo sapiens have descended from the Afro-Asian Cro-Magnons.

Second (and this is really, really important!) - Darwin called for *"innumerable intermediate forms"* to appear in the fossil record. In 1859, as in today, there is *a complete absence of these transitional forms.* Of all the fossil discoveries, none have been able to substantiate Darwin's theory linking ape and man - for 115 years. *Zero!* Small sharks have evolved into larger sharks, no shark has ever evolved into an amphibian. Some evidence of the lack of these forms as follows:

- biologists have searched for one bacteria blending into another, but no blend exists
- Darwin's evidence of the finch evolution resulted in three different varieties of finches, but no new bird species
- the fossil record actually reveals that every class, order, genus or species simply *appears*, it has not trans mutated into another

Even in 1871 Darwin's biggest critic St George said the following -

"Natural selection does not harmonize with the coexistence of closely similar structures of diverse origin. Certain specific differences are found to have appeared suddenly rather than gradually, and there are many remarkable phenomena in organic forms upon which Natural Selection throws no light whatever."

In *Darwin's Black Box*, Michael Behe demonstrates that Darwinian evolution is unworkable at the molecular level. That the history of biology suggests a series of black box evolutionary processes, that stay within certain boundaries and borders. It does not cross from one line to another, like apes to humans. Random mutations and natural selection cannot create humans from apes.

Neanderthals appeared around 200,000 years ago, peaked out at 75,000, declined at 35,000, then became extinct at 30,000. Cro-Magnons, however, thrived.

Intelligent Design hypothesizes that an "unevolved intelligence" embodies the true origins of man. Some scientists believe that Intelligent Design works with a limited form of evolution, others believe that Intelligent Design denies any involvement of (Darwinian) evolution. But, according to William A. Dembski and Michael Ruse - "...these disagreements are minor compared to the shared belief that we must accept that nature operating by material mechanisms and governed by unbroken natural laws, is not enough...The problem is that natural selection cannot account for its own success" ("Debating Design" - Dembski and Ruse).

In Darwin's Black Box, Michael Behe "argued that the irreducible complexity of certain biochemical systems convincingly confirms their actual design" (Behe).

"In the light of the information gathered thus far, it would appear that in so many instances anthropologists have, for the past century, been making a rigorous study of ancestors that we never had. For the most part, they've been recording generations of prehistoric Apes that were actually the ancestors of today's apes, and we're nothing whatever to do with eventual humankind." (Gardner - pg. 75)

CHAPTER 7

ᚱELIGION?

⸘𒀀𒁹𒀭𒌋𒁹𒈨𒁹?

"Religion is regarded by the common people as true, by the wise as false, and by the rulers as useful." - Seneca the Younger - Roman Philosopher

"Surely the ass who invented the first religion ought to be the first damned". – Mark Twain

"The world holds two classes of men – intelligent men without religion and religious men without intelligence" – Abu Ala Al-Maari (Middle East poet)

"With all this continual prayer, why no result? (Religion) wholly misrepresents the origins of man and the cosmos...Literature, not scripture, sustains the mind"

"We Believe with certainty that an ethical life can be lived without religion, and we know for a fact that the corollary holds true - that religion has caused innumerable people not just to conduct themselves no better than others, but to award themselves permission to behave in ways that would make a brothel keeper, or an ethnic cleanser raise an eyebrow."

"Religion has run out of justifications. Thanks to the telescope and the microscope, it no longer offers an explanation of anything important. Where once it used to be able, by its total command of a worldview, to prevent the emergence of rivals, it can now only impede and retard - or try to turn back - the measurable advances that we have made." - Christopher Hitchens - *"God is not Great"*

RELIGION ?

WHY IS ORGANIZED RELIGION on the decline worldwide? Does organized religion actually promote enlightenment, or is it inevitably a control mechanism of enslavement? Many religions today breed hypocrisy and intolerance. Shame, guilt, and sin are the benchmarks of many of our religions. The division among religions is the cause for many wars worldwide. Many religions are completely out of touch with modern society. Almost 10% of the United States are atheist or agnostics.

There are several descriptions within the Bible that suggest UFO's and the Anunnaki-

When Moses spoke to God on the burning bush atop Mt. Sinai, why did his face "glow" so much so that his people could not look at him, and he was forced to wear a veil? Was he burned? Can't it be suggested that he was irradiated? Was he suffering from microwave exposure? Couldn't the burning bush be construed as a UFO? I believe that Moses entered into a UFO atop that mountain and was in fact exposed to radiation.

When Ezekiel had a dream, and took a "Stairway to Heaven", couldn't this also be construed as having entered a UFO? The following is a direct quote from the Bible in Ezekiel, The Glory of the Lord -

"As I looked, behold, a stormy wind came out of the north, and a great cloud, with brightness around it, and a fire flashing forth continually, and in the midst of the fire, as it were gleaming metal"

Gleaming metal? Where do we find *gleaming metal* in 2000 BC? Isn't this a direct reference to a UFO landing? Even "fire flashing forth" can be construed as the exhaust from its propulsion, couldn't it? Read further...

"And above the expanse over their heads there was the likeness of a throne, in appearance like sapphire; and seated above the likeness of a throne was a likeness with a **human** *appearance"*

Now wait a minute.....did Ezekiel just say, "a human appearance"?? How is that possible? How can a human-like being be "above their heads" and sitting on a "throne"? Firstly, if the appearance was human, then it couldn't be an ethereal God. Secondly, if it was a throne, then how can an ethereal God be sitting on it? Now, you could argue that it was the "likeness" of a throne and not a real throne.....and you'd be right....but what would an ethereal God need a throne for anyway? Why would an ethereal God need to sit on a chair? In my opinion, Ezekiel saw a human being actually seated on a high-backed chair in a UFO.

A bigger questions begs - if God created man "in his own image", then how can "God" then be ethereal? God from the Bible must then be human-like! I believe in the concept of the big "G" and the little "g". The big "G" is the ethereal, non-corporeal entity that is the creator of all things (including the Anunnaki), whereas the little "g" from the Bible represents the plural god-like humanoids that created us.

1. God did not intervene during the Holocaust when his "Chosen People" were being murdered by the millions

2. The Sumerian "Atra Hasis" was written more than a thousand years before the Old Testament

3. Enlil told the Jews that he was God – but he lied- he also lied when he told Adam and Eve that they would "surely die" if they ate the fruit from the Forbidden Tree, but they didn't die.

4. We were created by the Anunnaki, used as slaves to mine gold, then we were nuked when we either "got too smart" or overpopulated.

5. We have been conditioned, and our spiritual growth has been stunted as a result of religion

6. The "Missing Link" only exists because we were genetically manipulated by the Anunnaki. There is no link, because the link was CHANGED by humanoids

7. The "truth" that we were educated to believe starts very young and becomes ingrained in our psyche. We grow up in a religion, believe it to be the only "correct" one for some reason or

another, marry into that religion, bring up our children according to the values of that religion, associate with people within that religion, etc. etc. For what? For who? For God?

8. "Worship" is translated in ancient times as "to work for", not to "praise".

9. "The Brotherhood of the Snake" are followers of Enki that strive to eliminate the "bondage " of the human race from the "pretender gods" like Enlil

10. The word "Elohim" from the Old Testament is a plural term, signifying gods or an assembly of gods

11. The Dead Sea in Israel is dead because of a nuclear explosion. Evidence is found in the physical creation of a tongue-shaped sand barrier called El-Lissan, from which the waters from the pre-existing salt sea had been nuked.

The Tower of Babel Story according to Jack Barranger in *"Past Shock"* doesn't make sense - *"According to Sitchin, the above is a much more accurate accounting of the Tower of Babel story than the tale of a group of "gone astray" humans who were simply trying to build a high tower in order to reach God. This is the commonly accepted Sunday school version. If God was already there in their presence, why did these humans feel a need to build a tower to get closer to God? This just doesn't compute"* – Past Shock – Barranger pg. 43

He continues – *"Our creators were emotionally depraved and spiritually bankrupt. They saw the rapidly increasing intelligence of their creations as a threat. Then "the Lord" in all his "wisdom" and "compassion" creates such confusion that these humans can no longer communicate among themselves. This is recorded history's first intentional dumbing down"* (pg. 44)

Here's an interesting quote from the Torah (Hebrew Bible) –

"The Nefilim were upon the Earth, in those days and thereafter too, when the sons of the gods cohabited with the daughters of the Adam, and they bore children unto them. They were the mighty ones of eternity – the people of the shem."

Notice the use of the phrase *"when the sons of the gods"* – plural. Not God, or the son of God – son_s_of the god_s_. There's a reason for this....the Torah is speaking about the Anunnaki (gods) and the Nephilim (angels, or lower level ET's). I find it very interesting that they are referred to as – *"the people of the shem"* for two reasons. Hashem – is the name of the Hebrew god according to the Torah. A "shem" in Sumerian language means a "rocket". So, I take Ha -shem (the Hebrew God) to literally refer to the Nephilim in the rockets. So, this verse basically means that the Nephilim (sons of the gods), came down in their rocket ships and had sex with the female humans of earth, who had their babies.

CHAPTER 8

THE BOOK AND
THE MISSION

THE COUNCIL OF NINE

Ε͡ΙΙ R. ANDRIJA PUHARICH, an American doctor born in Chicago of Yugoslavian descent, claims to have paranormal abilities, Dr. Puharich is Uri Geller's (renowned psychic) mentor. Dr. Puharich claims to have summoned an extraterrestrial who had identified himself as "M". "M" claims to be a member of the Council of Nine. Nine Principles and Forces of Nature. M said to Puharich-

"God is nobody else then we together, the Nine Principles of God. There is no God other than what we are together"- "The Stargate Conspiracy" – Picknett and Prince

The Nine claims to have left a 'Knowledge Book' in Egypt (under the paws of the Sphinx? – author's comment) 6,000 years ago. They claim to be the 'Elohim' from the Old Testament, and to have originated from the planet Sirius. (see comments regarding Robert Temple's book – *The Sirius Mystery*)

The Tower of Babel is an attempt by us to build a rocket-ship, not just a tall building. We were trying to mimic a rocket ship, similar to the ones used by the Anunnaki, or what's referred in the Bible as a "shem". We wanted to travel to the heavens to be like our makers. But having discovered this mock spaceship, the Nephilim were so angered that they said – "Now, anything which they shall scheme to do shall no longer be impossible for them". That is when the Nephilim confounded our language so that we were unable to cooperate and complete the "rocket ship".

Ezekiel, a character from the Bible, encountered 4 humanoid aliens and went aboard their ship. From EZEK 1:5 - 1:16 - "...*also out of the Mist there comes the likeness of four living creatures. And this was their appearance they had the likeness of a man. And the living creatures ran and returned as the appearance of a flash of lightning. The appearance of the wheels and their works was like unto the color of a beryl stone and they four had one likeness and their appearance and their works was as if it were a wheel in the middle of a wheel.*" The depiction of a "wheel in the middle of a wheel" suggests a UFO. The color of beryl is aluminum, and it is of great hardness occurring in hexagonal prisms. How can this be anything else but the description of 4 extraterrestrials coming down in a UFO?

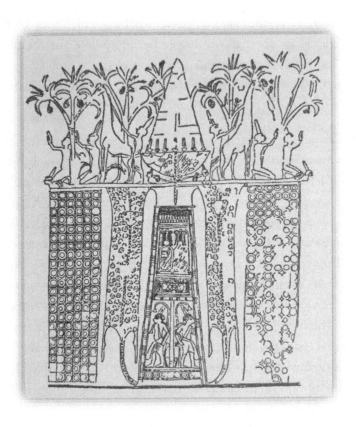

SUMERIAN DEPICTION OF A ROCKETSHIP

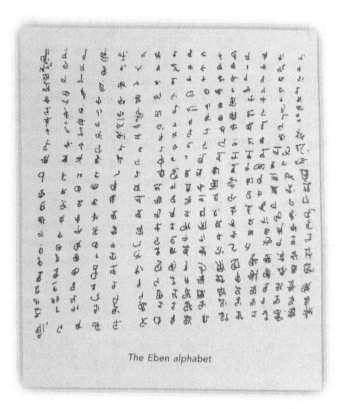

The Eben alphabet

THE EBEN (GREY ALIENS FROM ZETA
RETICULI STAR SYSTEM) ALPHABET. THE
ROSWELL ALIENS WERE FROM THIS PLANET,
AND THIS WAS THEIR ALPHABET.

THE YELLOW BOOK ■

Also called the "Orion Cube" was provided to us by the Greys in April 1964. It is essentially a transparent book 6.35cm thick constructed of a clear, heavy fiberglass- type material. The border of this book is a bright yellow color, hence the name "The Yellow Book". It appears to be a holographic image generator in which 3D pictures emerge when the reader looks upon the transparent surface. The book is read based upon the thought and language of the particular reader. Depending on the particular language the viewer is thinking, that particular language will appear. To date, scientists have identified 80 different languages. The book appears to be the true and correct complete history of humankind as portrayed by the EBEs. (Extraterrestrial Biological Entity) The book was used during the Montauk experiments and the Philadelphia Experiment.

THE RED BOOK ■

The Red Book is a detailed account written and compiled by the US government on UFOs dating from 1947 to the present. It is an orange- brownish book that is updated every 5 years. It contains volumes upon volumes of information that government agents have gathered regarding our interactions with a dozen or more extraterrestrial biological entities. The Red Book is presented to the sitting president of the United States every 5 years.

THE MISSION TO SERPO

N APRIL 1964, THE FIRST pre-orchestrated communication occurred at Holloman Air Force Base in New Mexico.

On July, 16 1965, the US sent 12 astronauts from the Groom Lake complex in Nevada to planet Serpo, 38.42 light years away. This project was directed and monitored by the DIA. The project was called "Crystal Knight". We discovered on their planet that they had not developed atomic energy! However, they did develop very powerful particle-beam weapons. They landed here on Earth on April 24, 1964. Serpo is in the Zeta Reticuli Star System. NASA had no involvement at all in the Serpo Mission. 12 astronauts went on the mission and were identified by 3-digit numbers. They consisted of-

Team Commander, Assistant Team Commander, 2 Pilots, 2 Linguists, a Biologist, 2 Scientists, 2 Doctors and a Security Person

The film and audio recordings of the trip are stored in a vault at Bolling Air Force Base in Washington, DC.

Serpo is 3 Billion years old (younger than Earth). Their two suns are 5 Billion years old. The Eben civilization is estimated to be only 10,000 years old. They had to relocate to Serpo after 5,000 years because of problems involving extreme volcanic activity in their original home planet. Serpo contains carbon, hydrogen, oxygen and nitrogen. Zeta Reticuli is only 37 light years away from Earth.

Serpo had numerous volcanoes, mountains at an elevation of 15,000 feet, snow at a maximum depth of 20 feet around its Northern Pole, lush green fields of grass, evidence of earthquakes, animals – the team found one similar to an armadillo that was hostile. One similar to an Ox – this one was timid. A mountain lion and a snake with "human-like eyes". Serpo contained a body of water that did not contain fish, only a small eel 10" in length. Two types of flying creatures – one similar to a hawk, and another similar to a flying squirrel. Insects similar to cockroaches but smaller.

The Team returned to Earth on August 18, 1978, but only 7 returned. Two members decided to stay on Serpo, and three died- one en route to the planet and two on the planet. The last survivor of the trip died in 2002 in Florida.

The climate in Serpo was very hot – in excess of 130 degrees Fahrenheit. The Eben food was tasteless and gave the humans gastrointestinal problems. Serpo had two suns, so complete darkness didn't exist there. The Ebens did worship a God. The population of the planet was roughly 650,000. The Ebens were vegetarians.

The journey to Serpo took roughly nine months to travel the 38 light years. The Team stayed on Serpo for thirteen years. The Team returned to Earth in an Eben spacecraft which only required seven months for the return.

It has been documented that a treaty with extraterrestrials was entered into in 1954 by President Eisenhower called the "Grenada Treaty" which was effectively an exchange of animal blood (cattle mutilations) for military weaponry and intelligence (anti-gravity, lasers and chips). Laser weapons called "Armorlux" were exchanged. This exchange was dubbed – "Project Plato".

THE ONLY KNOWN PHOTO TAKEN FROM
PLANET NIBIRU

ET TYPES

 TYPES OF ET'S have been documented, however 57 different races are believed to exist. Some are "good", support the evolution of humans, and some "bad" who support de-evolution. The bad aliens sap life, need blood from living organisms, possess mind control, implant devices, and use photosynthetic digestion.

1. GREYS also called Ebens - EBE-1, and EBE-2 helped us reverse engineer craft since 1953- they are from Zeta Reticuli, Orion and Bellatrix—friendly- offered diplomatic exchange program. Assisted in our nuclear program. Addressed "Biorhythms" of Earth every 40 years - 8/12/43, 8/12/83 and the next on 8/12/23. They visit Earth often are non-violent and are like worker bees. We have information on 22 types of greys but have been visited by only 2.

   a.   *Short Grays* (4' tall) - from Zeta Reticuli - (EBE-1) - 1500 light years away

   b.   *Tall Grays* (7' tall) - (EBE-2) - from Rigel, Betelgeuse and Alpha Centauri A

2.   NORDICS from Aldebaran

3.   REPTILIANS (also called Draconians) - from Orion, Alpha Draconis and Sirius B. Some Reptilians are humanoid in appearance. Elite Draco's have wings and tails and are violent in nature. They are cold blooded with greenish blood, do not produce sweat, hibernate and are capable of space travel.

5.   TALL WHITES

6.   SIRIANS from Sirius B and Sirius A

7.   VEGANS from Constellation of Lyra

8.   PLEIADIANS from Erra in the Pleiades- rode "beam ships". The "K Group" or Kondrashkins supposedly met with FDR in 1938 to create "Psychor" - A character named Emil P. DeCostain was apparently a member of the K Group.

9. TAU CETIANS from Tau Ceti

10. ANDROMEDANS from Andromeda Galaxy.

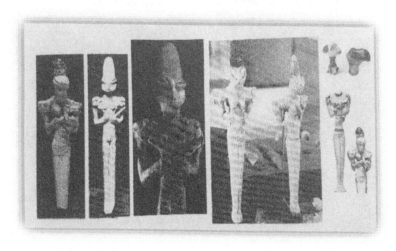

CARVINGS OF ALIEN BEINGS FOUND IN
SUMER. REPTILIANS?

FAMOUS PHOTO OF THE ROSWELL CRASH
DEBRIS. A PARACHUTE? OR ALIEN
MATERIAL?

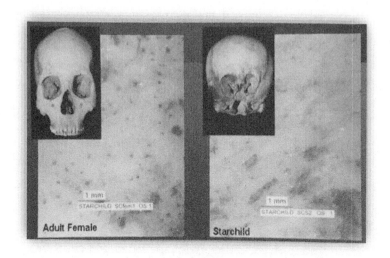

COMPARISON OF HUMAN SKULL WITH THE
"STARCHILD SKULL" OF EXTRATERRESTRIAL
ORIGIN FOUND ON EARTH.

DECEASED AUTHOR LLOYD PYE
DEMOSTRATING THE STARCHILD SKULL.

Alex Collier and Billy Meier have suggested that all human life in the galaxy originated in the constellation Lyra, which is in the vicinity of the Ring Nebula or M57. Making the original group of human extraterrestrials "Lyrans".

Our creators, the Anunnaki are approximately 350,000 years ahead of us. When they bio-engineered Adam, they did not give him the gift of what the Bible describes as "to know". What does "to know" mean? Most people may speculate that "to know" means to have sexual relations with. While that is close, it is actually the ability to "procreate" or "reproduce". When we were first created we had very little self-consciousness and we were sterile. Our makers, the Nephilim (the sons of Anunnaki) only wanted a servile homo erectus without complications, only maximum output and sufficient intelligence to work the mines for gold. But eventually, our "model" was improved, because of their necessity for more and more workers, to procreate ourselves! What a wonderful gift given to us out of pure need by our makers!

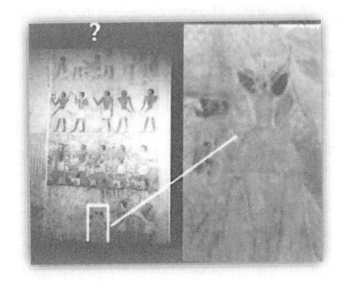

EGYPTIAL PAINTING OF AN EXTRATERRESTRIAL
CIRCA 2000 BC

CHAPTER 9

THE MAN BEHIND
THE CURTAIN

WE THE PEOPLE

In Uncle Sam we trust...

LOVE AMERICA - Everything about it. A land of immigrants, live free in a capitalistic society. Free to think, speak, worship, develop any business you like (as long as it's legal). In exchange for your freedom, you pay taxes. You trust in your government and your leaders. You are happy.

But here's the problem - your own government is not straight with you. Not straight about the extraterrestrial contacts, the environmental situation, and cures for many diseases including cancer. The government is run by the super-wealthy who will do anything to control you and protect their financial interests at the expense of your own. There really is no reciprocation. We are being callously exploited and hoodwinked. Only those who have the courage to think critically will see through this scheme. So much has been hidden from us, both information and truth in the form of evidence.

Brad Olsen, the author of 3 books on Esotericism, expresses this beautifully -

"It is a historic and highly effective propaganda technique used also by the church in the Middle Ages as well as by all greedy power structures before and since. Their goal is to maintain power and control an astonishingly destructive addiction. We've been lied to and conditioned beyond belief, so subtly and subliminally has information been manufactured, produced then forced upon us. We hardly even realize it because the perpetrators are those very people and institutions we have long relied upon. This kind of deceit is so expansive and huge and its size and subtlety that it surely is too unbelievable to be true." - "Modern Esoteric"

One thinks about Nikola Tesla and the threat he posed to the wealthy industrialists who controlled energy. His Free Energy machines would effectively put them out of business, so they pulled all of their investments out of Wardenclyffe and instead shut him down. Olsen continues -

"Thus, in a way, when the populist holds a strong conviction for uncovering the truth, they have nothing left to lose. We then become dangerous people to The Establishment. The government fears minds such as ours, for we are no longer subject to State vices or control. They can no longer pull the wool over our eyes, can no longer seduce us with comfort, and no longer subdue us with Force." - " Modern Esoteric"

We are living through a process of indoctrination that has been handed down for centuries, of supposed truths perpetuated by the wealthy. It is a self-perpetuating system constructed to make you believe that it is organized for your own well-being, while in actuality, it is solely for their own self-interests. It is a "Service to Self" indoctrination, not a "Service to Others". It is patterned to fit the narrow and selfish interests of the wealthy.

Why don't our history books discuss the Sumerian tablets which date back civilization to 10,000 BC? Why don't they discuss "fallen angels" and "gods" written about in stone on them? Why don't they cover the magnificence of the pyramids and how modern society today cannot even begin to construct them? Why don't they talk more about Tesla and less about Edison?

*"The UFO situation is the most important subject in the history of the human race. The people not only have a right to know, **damn it, they need to know**. I'm angry because I see the constitutional system being rubbished, the cover up, the lies, the secrecy, the 'Black Budget' expenditure of $50 billion a year of taxpayers' money, spent by people who are completely unaccountable. This is all a total violation of the constitutional system…"* - Colonel Robert Dean US Army Command Sergeant Major with "cosmic top secret" clearance.

"The 21st Century will die laughing at the Condon Report" - John Northrup, founder of Lockheed Martin, addressing the official government report on UFO's

The ACIO (Alien Contact Intelligence Organization) is an official government agency dealing with extraterrestrial contact. Members of the ACIO hold the highest level of security clearance in the United States called 'Cosmic'. No US president has ever been cleared for this level of security and only 25 people in the history of the world have ever reached this level. The ACIO Deals with UFOs, aliens and 'Particalization' - the principles behind time travel.

Al Bielek is a whistle-blower that worked for the government during the Philadelphia Experiment aboard the USS Eldridge. Both Albert Einstein and Nikola Tesla assisted in creating a portal using Unified Field Theory and String Theory to render the ship "invisible" and transport it from one location to another. Photos supplied by Morris K. Jessup supports and proves that the Philadelphia Experiment was an actual event and was successful! Tesla is documented to have created "sensitive transmitters" in order to communicate with extraterrestrials during this experiment.

I discussed the Roswell crash in my previous book - **"Adam = Alien"** - and the fact that the government's attempt at covering it up was a failure. Walter Haut, who was First Lieutenant of the 509th Bomb Group at Roswell Air Field, broadcast to the world the following quote-

"The many rumors regarding the flying discs became a reality yesterday when the intelligence office of the 509th Bomb Group of the Eighth Air Force, Roswell Army Field, was fortunate enough to gain possession of a disc through the cooperation of one of the local ranchers and the Sheriff's office of Chaves County. The flying object landed on a ranch near Roswell sometime last week. Not having phone facilities, the rancher stored the disc until such

time as he was able to contact the Sheriff's office, who in turn notified Major Jesse A. Marcel of the 509th Bomb Intelligence Office."

This broadcast was simultaneously interrupted by a radio transmission from higher headquarters, saying:

"ATTENTION ALBUQUERQUE: CEASE TRANSMISSION. REPEAT. CEASE TRANSMISSION. NATIONAL SECURITY ITEM. DO NOT TRANSMIT. STAND BY...."

If this really was a weather balloon, why the interrupted *national security transmission* to cease the message?

Stanton Friedman, whom I had met at a recent conference actually found the actual agent that wrote the memo, who told Stanton - *"Happy in my retirement. No guys in black suits on my doorstep. I cannot talk to you".*

Mac Brazel was the rancher that found the disc, and when asked by KGFL Roswell reporter Frank Joyce about the incident, Brazel said - *"Frank, you know how they talk of little green men?... They weren't green".*

US botanist Guillermo Mendoza studied EBE-1 in 1951 and describes its photosynthetic digestive system, under the direction of President Truman and Project Grudge.

Some characteristics of the ET's were described by a nurse during the examination as follows:

o Their arms are longer from wrist to elbow than from elbow to shoulder

- They have four fingers with suction cups for tips
- They have no thumbs
- Their heads were pliable
- They did not have teeth, only something resembling rawhide
- Their skin was pinkish-gray, tough and leathery
- They had no red blood, only a colorless liquid
- They had no digestive system, intestines or a rectal area

REAGAN BRIEFINGS

1981- 54 audio-cassette tapes, were declassified in '07.

The Caretaker – "Mr. President, as was mentioned earlier, I must say, that this briefing has the highest classification within the US Government. I will start with a slide presentation. I have most of this briefing on the slides, but also I have an outline that I have passed out to each person in attendance."

President Reagan- "What does that mean? Do we have codes or a special terminology for this?

The Caretaker – "Mr. President, EBE means "Extraterrestrial Biological Entity". It was a code designated to this creature by the US Army back in those days. This creature was not human, and we had to decide on a term for it. So, scientists designated the creature as EBE1. We also referred to it as "Noah". There was different terminology used by various aspects of the US Military and Intelligence community back then".

A device was found at the crash at Los Alamos. It was thought that the device was meant for sending and receiving messages back to the ET home. But the scientists couldn't figure out how to work it for years. Finally, after using the energy source from the craft itself, did the device work.

The Reticulan EBEs gave the US government 500 lbs. of Element 115. It was in the form of discs (author – Dropa discs?). It was a reddish-orange super-heavy metal.

According to the Andromedans, there are over 135 billion human beings in our galaxy, and that the universe is a 21 trillion-year-old hologram. That this hologram contains 11 dimensions or densities. We Earth humans occupy the 3rd density. They claim that our human genome is the result of 22 extraterrestrial races tinkering with our DNA to produce us. According to the Andromedans, our Moon was transported here only 12,000 years ago from a planetary system in Ursa Minor and placed in its present orbit. They also claim that it is hollow (which was proven on our missions there)– astronauts heard vibrations on the Moon. There are remnants of 9 domed cities on the Moon that have been discovered by NASA and Russian astronauts. Skeletal remains of humans and reptilians were discovered on the Moon. The Andromedans claim that life exists on 7 planets and 15 moons in our solar system. Also, that there are over 100 trillion galaxies and 100 billion suns.

The Andromedans claim that there have been 3 nuclear wars on Earth in the last 450,000 years the most recent of which happened in 11,913 B.C. that our skin has changed from green to red (Native Americans, Egyptians and Mayans), to yellow (Asians), to black (Africans) and finally to white.

GOD BLESS THE UNITED STATES OF AMERICA

In my previous book – ***"Adam = Alien"***, I touched upon secret groups such as the Bilderbergs, who control the majority of the wealth worldwide. The Federal Reserve Bank in the US has postured itself as a "federal institution", reporting to the Federal Branch of the United States – although nothing could be further from the truth. It is a private institution, or group of international bankers. Based upon financial crises and chronic institutional, and personal insolvencies in the early 1900's – the US Government acquiesced control to "The Fed". Since 1913, the Fed introduced the dollar bill as a Federal Reserve note as "legal tender", approved by the US Treasury.

Isn't it interesting that two of our beloved presidents – Lincoln and Kennedy – who both tried to reinstitute power to the US Treasury, and away from private banking groups – were assassinated?

In 1862, Lincoln introduced the First Legal Tender Act which authorized the printing of "greenbacks" by the US Treasury that weren't backed by gold or silver. After which the US economy grew at a remarkable rate, uncontrolled by private bankers, and untaxed by foreign banks. In early 1961, Kennedy halted all sales of silver from the US Treasury and authorized the printing of "silver certificates". This created an American currency outside of the control of the Fed.

The Rothschild and Rockefeller families control the Fed. They formed the Bank for International Settlements, or BIS, creating the first worldwide central bank. Author David Wilcock describes its intricate deception as follows:

> *"On September 19, 2011 a Swiss scientific study led by Dr. James Glattfelder proved that a staggering 80% of all the money that was being made in the world was filtering back into the pockets of the Federal Reserve through a very carefully disguised 'interlocking directorates' of corporations.....Supercomputers were used to analyze a database of the top 37 million corporations and individual investors worldwide. Shockingly only 737 corporations controlled a network that was earning 80% of the world's profits. With even greater number crunching, this web of ownership could be further narrowed down to a super entity of only 147 companies. An astonishing 75% of the corporations within this super entity are financial institutions. The top 25 financial institutions within this highly covert group include – Barclays, JP Morgan Chase, Merrill Lynch, UBS, Bank of New York, Deutsche Bank, Goldman Sachs, Morgan Stanley, and Bank of America – all of which are allegedly members of the Federal Reserve. "*

Wilcock goes on to say – *"Representative Alan Grayson, former representative Ron Paul, and now deceased senator Robert Byrd forced through a congressional audit of the Federal Reserve in 2011 and found that the Fed secretly gave away $26 trillion worth of*

American taxpayers' money." 26 trillion!! This $26 trillion was used to bail out the top Federal Reserve banks themselves.

My point is discussing the Fed and the wealthy families like the Bilderbergs, the Trilateral Commission and the Council for Foreign Relations, is to demonstrate to you, the open-minded thinker, that the world is not as it appears to be. What we were taught in school, was not the (whole) truth. That aliens and UFO's do exist. They have existing a whole lot longer than us. It's because of them that we are here. These groups *have everything to lose* if John Q. Public knows the truth.....for thousands of years, the door has been shut tight. Those behind the door know the truth, are hiding the truth, and want John Q. Public to remain as dumb as possible.

The 'Club of Rome' is a group of 100 people that are leaders in finance, politics, the judiciary branch and big business. This club has been pledged to a consortium which controls all international finance. Their objective is to institute a world dictatorship - a "New World Order" - as in the words of both George Bush Sr. and Jr. A member of the Rockefeller Foundation in New York City has supervised the construction of the Parliament Building to accommodate the new World Government in Canberra, Australia.

The Tri-Lateral Commission gets its name from an insignia from alien craft. It is the letter "T" inside of a black triangle with a red background. Members of the

Delta Group - a force that provides security for all Above Top Secret programs have badges with this same insignia. Since the Greek letter delta is the fourth letter of the Greek alphabet, this refers to Planet Earth - the fourth planet from the sun. Masonic tradition also uses a triangle in several of its insignias. At the Dulce base in New Mexico, the insignia is the Greek letter tau (T) inside of an inverted triangle. Each US base involved in alien activity has a Greek letter inside of a triangle.

The CIA Act of 1949 made black budgets entirely legal. it states that -

"any government agency is authorized to transfer to or receive from the CIA such sums without regard to any provisions of law limiting to or prohibiting transfers between appropriations. Sums transferred to the CIA in accordance with this paragraph may be expended for the purposes and under the authority of sections 403a to 403s of this title without regard to limitations of appropriations from which transferred" (Morris - pg. 62)

CHAPTER 10

THE CODE
ADVANCES

STONEHENGE

I VISITED STONEHENGE IN 2014. It is an incredible site that is unmistakably built by an extraterrestrial race as a star calendar to track planetary movements in order to keep time and to predict seasons to prepare crops for growth. Seeing massive multi-ton stones built atop each other is impossible for humans to have achieved. The stones are too heavy, and the positioning of stones on top of other each other is a massive feat I believe was accomplished by levitation. Here are some facts surrounding Stonehenge that support my theory:

o   it is in the precise location of the Northern Hemisphere to track eight lunar observations
o   it predicts solar and lunar eclipses
o   80 bluestones were used (the hardest stones on Earth) which could not, and cannot be quarried by copper and/or bronze tools used in 2000 BC
o   these bluestones were transported 250 miles from the Prescelly Mountains
o   77 sarsen stones (hard sandstone) weighing 50 tons each, were moved 20 miles from Marlboro Downs
o   these sarsen stones were positioned upright, and as crossbeams atop each other

o the positioning of the bluestones and sarsen stones created such a perfect geometric pattern, not to create a tomb, but rather to create a sundial clock in order to predict a cycle of 18.61 years based on the sun and moon.

STONEHENGE

ATLANTIS OR ANTARCTICA?

According to Plato, the Egyptian priests fixed the date of the sinking of Atlantis at 9850BC.

DEPICTION OF ATLANTIS.

I agree with authors Rand and Rose Flem-Ath regarding Atlantis. It is not a sunken continent in the Atlantic Ocean, near Greece in The Mediterranean, or near Turkey or the Middle East. It is right beneath our noses. It is the island of Antarctica. Specifically, Lesser Antarctica. Here are the reasons:

1.   It is an island in the middle of the ocean, just as Plato said

2   It is "out of the regions of the South" and arised "after the deluge"

3.   Antarctica is 6500 feet in elevation- higher than Asia, South America, Africa, North America, Europe and Australia

4.   Was destroyed in 9600 BC, in the same century that agriculture sprouted. Coincidence?

5.   Were the Americas populated by Atlantians who fled after the deluge?

6.   The Piri Reis map showing Antarctica proves that it must have been mapped prior to glaciation- some 6000 years ago

7.   The Atlantis Blueprint's Appendix lists all the sacred sites on Earth (Giza Pyramids, etc.), which all lie on specific latitudes aligned with either the old or the new positions of the North Pole. In addition, the positioning utilizes the pi system of geometry to exact proportions, as if measured using advanced mathematics. How can the following famous landmarks possible fit an exact geodesic proportion to the Hudson Bay Pole, with the exception of a one-degree misalignment (which was as a result of shift of the Earth's crust prior to 9,600 BC?)

These sites all lie along the Earth's Energy Grid - Baalbek, Canterbury, Chichen Itza, Cuzco, Easter Island, Giza, Jerusalem, Lhasa, Luxor, Machu Picchu, Nazca, Newgrange, Quito, Rosslyn, Byblos, Jericho, Xi'An, Aguni, Pyongyang, Avebury, Abydos, and Nippur.

35,000 Sumerian tablets were found in Nippur, 1899 by American archaeologists. Dilmun was listed in the tablets as "the ancient Sumerian city dedicated to the flood god Enlil"- (Flem-Ath)

Dilmun was a mountainous island in the ocean. Most of its inhabitants drowned with the great flood. The survivors escaped in a great ship (the Ark?) and sailed to a mountain near Nippur. The tablets said that Dilmun, the island paradise from which they fled, lay "across the Indian Ocean towards the south – toward Antarctica"! – (Flem-Ath)

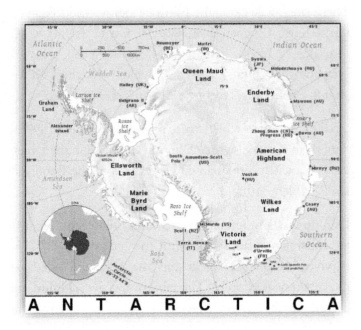

MAP OF ANTARCTICA WHICH FITS THE PLATO
DESCRIPTIONS OF "AN ISLAND" IN THE
"REGIONS OF THE SOUTH" IN THE "MIDDLE
OF THE OCEAN"

SIRIUS

The Dogon tribe, from Mali in Africa spoke of a third star in the constellation of Sirius which was invisible to the naked eye and by telescope.

1. The visitors from Sirius were amphibious, called 'Nommos'
2. The Great Pyramid represents Sirius B, and the Pyramid of Khephren represents the Sun
3. The Dogon knew that Sirius had an elliptical and not a circular orbit and was binary. How?
4. Sirius B is a "white dwarf" and contains a metal called sagala which is very heavy
5. Does 'Isis' come from 'Sirius'?
6. Anubis comes into play with the Masons – is the Great Sphinx's body actually that of Anubis (dog of knowledge) and not a lion?
7. Only 3 races that practice circumcision are – Colchians, Egyptians and Ethiopians
8. Was Oannes really Enki? If so, the original god of Noah was amphibious

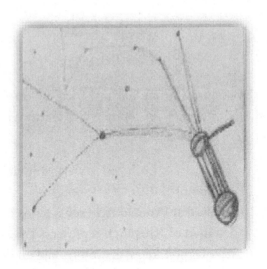

BETTY HILLS DRAWING - OF THE PLANETS
THAT HER ABDUCTORS WERE FROM. THIS
WAS DRAWN IMMEDIATELY AFTER HER
ABDUCTION.

SHE SAID THAT HER ABDUCTORS WERE FROM
THE PLANET SIRIUS. COMPARE TO THE
ACTUAL DRAWING OF THE SIRIUS STAR
SYSTEM NEXT ON THE NEXT PAGE.

COINCIDENCE?

THE ACTUAL DRAWING OF THE SIRIUS STAR
SYSTEM.

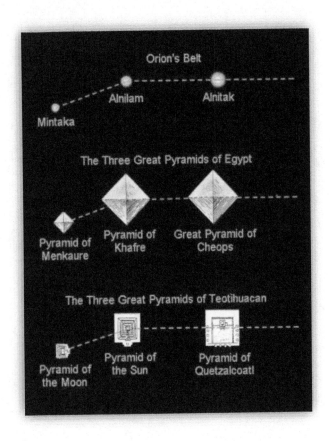

ORIONS BELT - PROOF THAT THE GIZA AND
TEOTIHUACAN PYRAMIDS WERE PLACED
EXACTLY AS A MIRROR IMAGE OF THE 3 STARS
OF ORION

MARS AND THE MOON

"It is a trick of light and shadows" - Dr. Arden Albee, California Institute of Technology commenting on the Cydonia Face

"NASA is not some Starship Enterprise on a mission to 'boldly go where no man has gone before'. On the contrary, NASA is the disturbed child of two dysfunctional parents - paranoia and war" - Graham Hancock (The Mars Mystery)

"The rocks are the first ever found on Mars that contain streambed gravels. The sizes and shapes of the gravels embedded in these conglomerate rocks- from the size of sand particles to the size of golf balls - enabled researchers to calculate the depth and speed of the *water that once flowed* across this location " - NASA Jet Propulsion Labs

N JULY 1976, THE VIKING ORBITER 1 took images of the Cydonia region on Mars. These images depict an enormous sculpture of a face as well as the ancient ruins of a great city. This city had been established thousands of years before the Earth was inhabited. This city was constructed by the Anunnaki using laser beams to cut stone and levitation to transport the stone that is used to erect them. One NASA photograph - PR95-17/HST.WFPC2 - it demonstrates that Mars has blue oceans and red landmasses.

Before he died, astronomer and mathematician Sir Fred Hoyle visited Dr. Gilbert Levin (head scientist at NASA that discovered life on Mars from the *Viking Lander* experiments) that "not only had the *Viking* detected life on Mars, but that, in the clandestine return sample mission to Mars, the U.S. government had obtained living microorganisms which were now under cultivation for potential applications mandated secret..." - Haze- Aliens in Ancient Egypt - pg. 203

The Elysium Quadrangle sector of Mars was photographed in 1972 - triangular pyramid-like structures were discovered. The *Mariner* photographs of B-Frames-MTVS 4205-3, DAS 07794853, MTVS 4296-24, and DAS 12985882 demonstrate that these photographs depict four actual, massive pyramid structures that cast shadows. These pyramids are larger than any known-on Earth. One of these pyramids is a five-sided structure called the D&M

Pyramid (named after its founders - DiPietro and Molenaar). These are not illusions.

The "Face" at Cydonia is real. It was first identified on *Viking* frame #35A72 by Dr. Tobias Owen. The Face, which shows an uncanny resemblance to that of the Sphinx (author's note: I believe it is the exact same face of the exact same maker - Ningishzidda). It measures a staggering 1.6 miles in length from crown to chin, 1.2 miles wide, and approximately 2,600 feet high! Teeth are apparent in the mouth, and the headdress resembles the same *"nemes"* headdress of the Ancient Egyptian pharaohs. The frame depicts enormous monuments as well.

Richard Hoagland, a former NASA consultant, and now author, in viewing frames 35A72 and 70A13, has identified additional monuments resembling a "Fort" and a "City".

A key point in proving that these structures are indeed artificial is that they are defined as "non-fractal" - or that their contours have been scanned and assessed as artificial and not natural.

The building block of life - water - is found all over Mars. Water spouts and steam vents appear. Several areas of terrain show areas of water erosion - similar to the erosion at the base of the Sphinx.

On one of Saturn's moons called Enceladus, there exists an icy veneer under which exists a large sea of water estimated to be the size of Lake Superior! David J. Stevenson, a professor of Planetary Science at the California Institute of Technology, claims that " *What we've done is put forth a strong case for an ocean*". Larry W. Esposito, a professor of Astrophysical and Planetary Science from the University of Colorado echoes this point - " *Definitely Enceladus because there's warm water right there now*".

Is the moon an artificial body? Was it made to incubate and foster life on Earth? Is it hollow? There are many questions that theorize that our Moon may not be of planetary origin but may have been literally *built* to enable life on Earth. The moon is bigger, older and lighter in mass than it should be, based on evidence of the characteristics of ordinary moons of other planets. When Apollo 13 crashed its booster onto the moon, the astronauts reported that it "rang like a bell", and "wobbled" as if it were hollow. The precise positioning of the moon and its gravitational effects on Earth, allow for a perfect condition to nurture life, unlike any other moons in our solar system. "Project Whiteout" is a US project that supposedly transports materials to the moon in effort to create a US base there.

The moon is depicted as an actual planet on Sumerian cylinder seals, based on its size and mass, which make it improbable that it was originally only a satellite of Earth, but rather a planet in its own right. The Sumerians actually *knew this*, from the Anunnaki.

THE DEAD SEA
SCROLLS

DEAD SEA SCROLL

TIMELINE

- ➤ 136 AD - Scrolls deposited at Qumran
- ➤ 1947 – Dead Sea Scrolls (DSS) discovered in caves
- ➤ 1952 – Copper Scrolls found, and 800 paper scrolls

Dead Sea Scrolls Content:

1. The location of 65 tons of silver and 26 tons of gold

2. An accurate inventory of the Temple of Jerusalem, and it's treasure!

The Dead Sea Scrolls Conspiracy:

The man appointed to manage, decipher and interpret the DSS was Roland de Vaux. De Vaux was a French priest born in 1903. He was anti-Semitic, referred to Israel (even after Israel's statehood in 1947) as Palestine, and knew of the explosive material contained in them. Several scrolls contained information not released to this day, that are so controversial, that the Vatican (and possibly the Israeli Government), refuses to translate and release it. This is information that some of the initial translators between 1950-1955 became embroiled in a literary "war of words" between each other and wrestled for control of the scrolls. De Vaux wanted only to release the data that supported the New Testament's ideologies. His mortal enemy was John Allegro, who worked with De Vaux translating the DSS for years. De Vaux used the Vatican and its power to silence Allegro and humiliate him. Allegro's intention was always pure – to release to the public *all* of the content of the DSS. De Vaux, on the other hand, had lots to hide. He released only 10% of the Scroll's content, and only the "bland" information contained in it. All the explosive material was muffled, and any attempt by Allegro to release it was met with ridicule and embarrassment.

At a conference held at New York University in May 1984, Professor Morton Smith commented – "I thought to speak on the scandals of the Dead Sea documents, but these proved too numerous, too familiar, and too disgusting. It observed that the international team (the Team captained by De Vaux) were 'governed' so far as can be ascertained, largely by convention, tradition, collegiality and inertia. The insiders, the scholars with the text assignments, 'the charmed circle', have the goodies – to drip out bit by bit. This gives them status, scholarly power and a wonderful ego trip." - "The Dead Sea Scrolls Deception" – pg. 72

Another scholar named Eisenman who also encountered barriers in attempting to challenge the International Team, in 1989, went public. He was quoted by The *New York Times*, the *Washington Post*, the *Los Angeles Times*, the *Chicago Tribune* and *Time* magazine, stressing five major points:

1. That all research on the Dead Sea Scrolls was being unfairly monopolized by a small enclave of scholars with vested interests and a biased orientation

2.   That only a small percentage of the Qumran material was finding its way into print and that most of it was still being withheld

3.   That it was misleading to claim that the bulk of the so-called 'biblical texts' had been released, because the most important material consisted of the so-called 'sectarian' texts – new texts, never seen before, with a great bearing on the history and religious life of the 1st century

In November 1990, the Israeli government appointed a Dead Sea Scroll scholar named Emmanuel Tov. The Rockefeller Museum and The Ecole Biblique became the headquarters of Qumran research. After Israel won the Six Day War, it came into possession of Arab East Jerusalem, and therefore gained possession of the Rockefeller Museum and the Biblique as "spoils of war".

Here's a summary of all the scrolls found-

1. **Copper Scroll** - found in cave 3, lists the location of a treasure of gold, silver and precious religious vessels. It dates from the time of the Roman invasion in AD 80.

2. **Community Rule** - found in cave 1, lists the rules and regulations governing life in the desert community. It establishes a hierarchy of authority, lists a "Master" of the community. Outlines punishments. Introduces the 'Messiah"

3. **War Scroll** - found in caves 1 and 4, is a war manual including strategy and tactics against the Romans.

4. **Temple Scroll** - found in cave 11, discussed the Temple of Jerusalem and its designs, furnishings and fixtures. It is similar to the Jewish Torah, comprising the five books of the Old Testament - Genesis, Exodus, Leviticus, Numbers and Deuteronomy

5. **Damascus Document** - found in the loft of an ancient synagogue in Cairo, an accumulation of worn-out religious texts. The texts were incomplete. However, the content of the texts was - *"provocative, potentially explosive "* (Baigent-pg. 145). It speaks of a sect of Jews who "remained true to the law". (Baigent - pg. 145) They entered into a Covenant with God, similar to the one cited buy the "Community Rule". Specifies three crimes - wealth, profanation of the Temple, and fornication (taking more than one wife).

It also designates three types - the "Liar" who defects from the community and becomes its enemy, the Star" who upholds the laws of the community, and the "Sceptre" who is essentially the Prince of the House of David.

6. **Habakkuk Commentary** - found in cave 1, discusses the "Teacher of Righteousness", or the Leader, and the "Wicked Priest", or the Adversary. Was the wicked priest actually Jonathan Maccabaeus (160BC) or his brother Simon?

Eisenman, after reviewing all the data, recognized that de Vaux had been too cavalier in his conclusions, and that Roth and Driver had been correct. The scrolls had been authored by the Essenes - first century pacifist Jews.

When the Dead Sea Scrolls were found, it's interesting that fragments of the book of Enoch were among the pages.

DEAD SEA SCROLL CAVES

DEAD SEA SCROLLS CAVES FROM A
DIFFERENT ANGLE

BONES

"GIANTS Traditionally this have been attributed to hydrocephalic deformation or artificial skull-boarding techniques, but as the number of these skulls have been found and studied increases, it is obvious to researchers that certain skulls are naturally oversized, and have increased cranial capacities that are not the result of disease or artificial manipulation" - Richard J. Dewhurst

Giants are described repeatedly in the Bible. Goliath is one of them, described as being approximately 10' 9" tall, muscular and gruesome. David is described as an ordinary human, of ordinary stature and frame. We know that the Watchers have been described as having intercourse with the *daughters of man.* Could the giants be the offspring of the Watchers, having the traits of their DNA mixed with ordinary humans? The only way giants could have existed on earth was through interbreeding with humans. A race of Eljo or Elyo is described in the book of Enoch as having roamed the earth - men of legend, men of mythology. These giants exist in modern mythology as the Greek gods - Hercules, Dionysus, Poseidon, etc. They all carried the Watchers DNA traits - tall, muscular, bearded - with unique weapons like Thor's hammer, or Poseidon's thunderbolt which acted like current 21st century weaponry. The giants had six fingers and six toes.

Many skeletons of giants have been recovered in the sandy Arabian desert, France, and even in the USA. These skeletons range from 9' - 23' tall! These skeletons are undoubtedly that of the Anunnaki offspring (Nephilim from the Bible) mating with humans throughout the years between the flood and 2000 BC. Bones have been discovered in Ohio that measure 8' -15' in height with - 6 fingers, double rows of teeth and red hair! The Smithsonian would (of course) confiscate all of these bones discovered in the Ohio Valley from the late 1800's to the late 1900's. Why confiscate them?

ACTUAL HORNED SKULL CIRCA 1000 BC

ACTUAL PHOTO OF GIGANTIC HUMAN BONES.
PROBABLY 15 FEET IN HEIGHT

ACTUAL PHOTO OF A GIGANTIC HUMAN
SKULL

THIS HUMAN SKELETON MEASURES TWENTY
FEET IN LENGTH

𒐎233𒐎

ACTUAL GIGANTIC SKULL FOUND IN THE
MIDDLE EAST

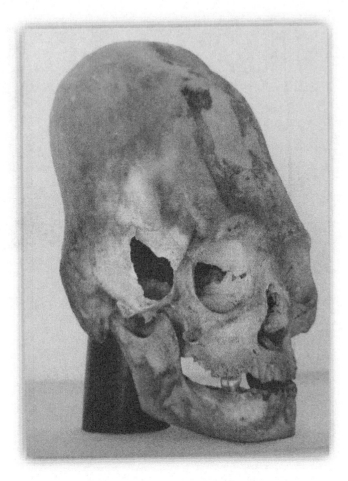

ACTUAL PHOTO OF SKULLS WITH ELONGATED
CRANIUMS FOUND IN PERU. THESE ARE NOT
SKULLS "BOUND" SINCE BIRTH FOR RITUAL,
THESE ARE UNTOUCHED ANUNNAKI SKULLS
THAT HELD A 2200CC BRAIN (COMPARED TO THE
CURRENT 1350CC BRAIN TODAY).

ANUNNAKI TREE

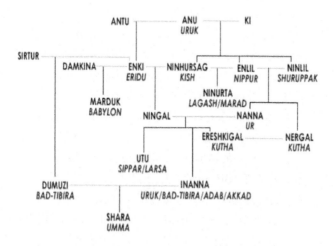

TOP = ANUNNAKI NAME, BOTTOM = CITY OF RULE

SUPPORTING DATA

1. NASA discovered water on the moon as far back as the Apollo missions. It took NASA over 40 years to confirm this. One of the first lunar rocks did contain rust. On March 7, 1971 lunar instruments detected an H2O vapor cloud on the moon that lasted 14 hours and covered 100 square miles. Soviets discovered water on the moon in the 1970's and published this fact in 1979.

2. General Nathan F. Twining, Commander of Air Material Group from the 1940's-1950's described the inside of the Roswell flying saucer as containing typewriter like keys that control the propulsion system, and a 35-ft long circular tube containing a clear substance and a copper-like coiling system inside of it.

3. According to Bruce Cathie, in his book - "Mathematics of the World Grid" - gravity and light are reciprocals of each other, and that UFO's manipulate these waveforms and frequencies in order to travel

4. Between 1946-1952, there have been 16 crashed UFO's containing 65 alien bodies.

5. Ronald Reagan's "Star Wars" Space Program that was conveyed as being a defense initiative versus the Russians was in reality, a defense initiative

versus aliens. It was introduced right after astronauts had taken 122 photos of alien bases on the moon.

6. The Greys were designed for space travel with flexible skin to cope with anti-gravity g-forces, had very slow metabolism, and a large heart that pumped a lymphatic-like fluid through their circulatory system. Their metabolism and aging were much slower than humans, their lungs had larger capacity and their bones were more flexible. They had difficulty breathing in our atmosphere. Their digestive system was similar to plants, having no waste product.

7. Alien craft "displace gravity through the propagation of magnetic waves controlled by shifting the magnetic poles around the craft so as to control, or vector, not a propulsion system but the repulsion force of like charges (electromagnetic anti-gravity propulsion)". - Piers Morris - "They Exist"

INDEX

BIBLIOGRAPHY

1. "Modern Esoteric" – Brad Olsen
2. "African Temples of the Anunnaki" – Michael Tellinger
3. "Alien Civilizations" - Will Hart
4. "Aliens in Ancient Egypt - Xaviant Haze
5. "Anunnaki Gods No More" – Sasha Lessin
6. "Architects of the Underworld" - Bruce Rux
7. "Atlantis Beneath The Ice" – Rand and Rose Flem-Ath
8. "Breaking the Godspell" – Neil Freer
9. "Debating Design" - William A. Dembski and Michael Ruse
10. "DNA of the Gods" - Chris Hardy
11. "Everything You Know Is Wrong" - Lloyd Pye
12. "FEMA Coffins and the Anunnaki" – Peter A. Grimm
13. "Flying Serpents and Dragons" - R.A. Boulay
14. "From Adam to Omega" – A.R. Roberts
15. "From Atlantis to the Sphinx" – Colin Wilson
16. "Future Esoteric" – Brad Olsen
17. "Gods of Eden" – Andrew Collins
18. "Gods of the New Millennium" - Alan Alford
19. "Gods, Genes and Consciousness" – Paul Von Ward
20. "Lost Race of the Giants" – Patrick Chouinard
21. "Past Esoteric" – Brad Olsen
22. "Past Shock" – Jack Barranger
23. "Return of the Golden Age" – Edward F. Malkowski
24. "Sacred Geometry and Spiritual Symbolism" – Donald B. Carroll

BIBLIOGRAPHY

25. "Secret Journey to Planet Serpo" – Len Kasten
26. "Space Faring Civilizations" – Eric Franz
27. "Sumerian Mythology" - Samuel Noah Kramer
28. "The 'Nonsense Papers" – James W. Astrada
29. "The Ancient Giants who ruled America" - Richard J. Dewhurst
30. "The Anunnaki Chronicles" - Zecharia Sitchin
31. "The Anunnaki Gods Return" - William King
32. "The Anunnaki of Nibiru" – Gerald Clark
33. "The Dead Sea Scrolls Deception" - Michael Baigent
34. "The Giza Power Plant" - Christopher Dunn
35. "The Grandest Deception" – Dr. Jack Pruett
36. "The Great Pyramid: A Factory for Mono-atomic Gold" – Spencer L. Cross
37. "The Lost Book of Enki" – Zecharia Sitchin
38. "The Mars Mystery" - Graham Hancock
39. "The Sirius Mystery" – Robert Temple
40. "The Son of God Who Walked the Earth" - Nathan R. Noble
41. "The Stargate Conspiracy" – Lynn Picknett and Clive Prince
42. "The Sumerian Controversy" - Dr. Heather Lynn
43. "The Synchronicity Key " - David Wilcock
44. "The Temple in Man" - R.A. Schwaller de Lubicz
45. "There Were Giants Upon the Earth" - Zecharia Sitchin
46. "They Exist" - Piers Morris
47. "What Egyptologists Don't Want You to See"- Jerret Gardner

AFTERWORD

PEOPLE HAVE OFTEN ASKED ME when I would be writing a follow up to *"Adam = Alien"*, and I'd always respond "soon, soon, probably next year". Shortly after *"Adam = Alien"* was published I received a flurry of invitations to speak at Colleges, Universities, UFO Conferences, even at local restaurants whose owners were interested in the subject matter. I sold my business of 25 years, changed occupations and then recently, lost my father who died at 94. He was a decorated WW11 veteran fighting in Europe from 1941-1945. Whenever I asked him if he'd ever seen or heard about Foo Fighters during the War, he surprisingly said, "No, not really".

A supremely humble man, who was frugal with words, was as honest as they come - a true gentleman. I couldn't believe that he had never seen or heard about Foo Fighters, when thousands of veterans swore by them. Maybe since he had been with a platoon of land-based amphibious engineers from Brooklyn, NY, they had not been exposed to airborne Foo Fighters like the Air Force pilots would have been. But you would think that he'd at least heard about them. "No, not really". I hear "no, not really" quite often when I ask people if they believe in UFO's, would consider alternative truths to the Bible, or believe that extraterrestrials have indeed landed here on Earth - even in the U.S! But it doesn't deter me. I keep arguing, fighting, coaxing, prodding away with more and more and more incontrovertible evidence to support my thesis (and others shared by so many notable authors in the Ancient Alien space). **I believe in it so strongly that it has become fact.** I believe in it so strongly that I changed 180 degrees from a religious, God-fearing boy to an agnostic non-God-fearing adult, to a secular atheist who only attends religious ceremonies out of respect for others' rights, beliefs and freedoms. I am now an atheist, who believes in a master architect in the universe, just not the one that the Bible purports to be the only God of the Bible. Yes, there is an elegant grand design, beautifully constructed through complex mathematics that pervades from the spiral helix of our DNA strands to the deepest abyss of a Black Hole.

If you, my dear reader, could believe just 10% of what I have written, then I have succeeded. I hope that I have enlightened you just bit, the same way that authors like Sitchin, Pye, Alford, and Von Daniken have truly inspired me. There are so many secrets out there. So many people with so many agendas. So many people who blindly follow other people's direction. So may people who truly believe that world governmental leaders have their best interest at heart. So many people who believe that only what they are taught in school is the truth.

So many people who believe every word in the Bible. Am I one of those people?

No, not really.

ABOUT THE AUTHOR

L EON BIBI is the author of *Adam = Alien Vol. 1.* Leon is a historian and avid researcher on all subjects regarding archaeology, human and ancient history, biology, Egyptology and religion. Leon has been a featured guest speaker on George Noory's "Coast to Coast" radio show and will be on the Ancient Aliens Series of the History Channel next year. He has been invited to speak about the Ancient Alien subject at various conferences, and both radio and television. Leon has a B.A from Washington University in St. Louis and attended Law School at the University of Miami (FL). He was a CEO of a Consumer Products company for 22 years, and a member of the prestigious "Young Presidents Organization" for 17 years. He is a professional drummer playing and recording music throughout his adult life. Leon lives with his family on the East Coast of the United States.

CODE CHALLENGE

BELOW IS THE CODE.
IT WILL UNLOCK THE 4000-YEAR OLD
SECRET.
THE CODE HAS NEVER BEEN DIVULGED
TO MODERN HUMAN CULTURE.
IF YOU CAN CRACK THE CODE, YOU WILL
LEARN THE MOST IMPORTANT SECRET IN
OUR LIFETIME.
A SECRET THAT THE MOST POWERFUL
MEN AND WOMEN IN THE WORLD WILL DO
ANYTHING TO PROTECT.
A SECRET THAT MUST BE REVEALED.

ONLY YOU HAVE THE

BOOK 1 OF THE ADAM TRILOGY

ADAM = ALIEN *Vol.1*

The first book of the ADAM trilogy by Leon Bibi. It explodes onto the scene with intricate detail about the origins of Adam. It explores the Anunnaki and their arrival here on Earth hundreds of thousands of years ago in search of gold to save their dying planet. Pyramids, Tesla, Propulsion, UFO's and extraterrestrials are examples as Bibi goes through his own transition from a skeptic to a believer. Some of the information in the book will shock you.

ON SALE NOW AT

www.amazon.com or
www.adamalien.com

BOOK 3 OF THE
ADAM TRILOGY

This book will delve into blood and chromosomal evidence of extraterrestrial tampering and intervention in the creation of modern day homo sapiens. It will explore the motives and actions of the Anunnaki gods who used their own DNA to create the world's first human hybrid hundreds of thousands of years ago. It will demonstrate that the supposed mythology of the Sumerian tablets which clearly told its story was a true record of our origins. How the Old Testament and New Testament was a recapitulation of the tablets yet altered to serve a purpose. This purpose was self-serving.

The truth lies in these tablets, spoken without an agenda, recounting the story of the creation of the Earth and all multicellular organisms created not solely through Survival of the Fittest and Natural Selection. We had help from the gods.

How did they do it? What secrets did they bestow upon us? What level of intelligence can we reach?

Book 3 will unveil all of this and more.

The Adam Code continues...

ADAM

DECŌDED

VOL. 2

BY

LEON BIBI

Made in the USA
Middletown, DE
20 January 2020